EDUCATION IN A COMPETITIVE AND GLOBALIZING WORLD

PROJECT BASED LEARNING ON ENGINEERING

FOUNDATIONS, APPLICATIONS AND CHALLENGES

Education in a Competitive and Globalizing World

Additional books in this series can be found on Nova's website under the Series tab.

Additional e-books in this series can be found on Nova's website under the e-book tab.

EDUCATION IN A COMPETITIVE AND GLOBALIZING WORLD

PROJECT BASED LEARNING ON ENGINEERING

FOUNDATIONS, APPLICATIONS AND CHALLENGES

JUAN VALENZUELA-VALDÉS
AND
PEDRO PARDO FERNÁNDEZ
EDITORS

nova publishers
New York

Library of Congress Cataloging-in-Publication Data

ISBN: 978-1-63482-217-6

Published by Nova Science Publishers, Inc. † New York

CONTENTS

PREFACE

Teaching of science and technology compared with those of other subjects always presented the handicap of the difficulty that confronted the student's understanding of theoretical concepts. Educational research has demonstrated repeatedly the shelter of the students in the manipulation of mathematical relationships, probably due to underlying conceptual difficulties.

Only a few students, with the exception of those with a clear theoretical formation, are those that pleasantly accept the task of understanding the mathematical formulation which set out the main scientific laws where other applied scientific and technical disciplines are based. This is the case for various branches of engineering, and often has been established an unbridgeable gap between the theoretical knowledge in the engineering and its application to solve problems.

In many cases, this barrier prevents the construction of knowledge thus resulting in watertight compartments in our brain; that is, theoretical knowledge of which the mathematical formulation can be understood, but not its ultimate meaning or the skills that help to solve problems. At this point is where the use of innovative teaching methods can help shorten that leap, especially in areas such as engineering, where traditionally not been taken into account this aspect. However, the main cognitive theories used in education, show that the basis of the education in Sciences, and perhaps of all teaching, should be to achieve a lasting understanding phenomenon, thus preventing the rote learning and trying, in their design to achieve a "meaningful learning".

Here, we present a work that develops a methodology based on a known technique based in solving problems, but adapted to an area where traditionally had not been used, such as engineering. It goes a further step by showing a concrete application which is based on projects in the context of communication and information learning technologies. These facts make this work of very interesting reading for both teachers and researchers in this field.

The following pages are the result of the work of an enthusiastic group with many years of dedication to educational research. It is a text to be read "in the same way that chickens drink water," as Prof. Tierno Galván said. If you do so, surely to you find many aspects that have sometimes overlooked.

María Isabel Suero López
Full professor
Physics Department - Science Faculty
University of Extremadura (Spain)

In: Project Based Learning on Engineering ISBN: 978-1-63482-217-6
Editors: J. Valenzuela-Valdés et al.

Chapter 1

INTRODUCTION TO PROJECT BASED LEARNING ON ENGINEERING

S. Román*[*,1], *B. Ledesma*[1], *A. Álvarez*[1], *F. Zamora*[2] and *E. Sabio*[1]

[1] Departamento de Física Aplicada, Universidad de Extremadura, Spain
[2]Departamento de Ingeniería Mecánica, Energética y de los Materiales. Universidad de Extremadura, Spain

ABSTRACT

The recent implantation of Bologna frame into University courses has led to a significant change in Teaching-Learning procedures, which are characterized by a more active participation of the student. The teachers can use different tools to enhance the success of this process; among them, Problem Based Learning (PBL) has proven to be an effective methodology. This work describes the main characteristics of PBL process, in terms of its features, practical implementation, evaluation and use in engineering disciplines. Moreover, an analysis about the experimental work made on this topic by several teachers in the University of Extremadura is made. Finally, the current state of PBL bibliographic sources in the different knowledge areas is accomplished, and the abundance of PBL scientific publications in the world geography is analysed.

[*] Emails: sroman@unex.es

1. Introduction

The recent adaptation of European countries to the EHEA (European Higher Education Area) in the frame of Bologna process has brought up significant modifications in the traditional Teaching-Learning strategies (TLS) [1]. As a whole, these changes intend to evolve the traditional teaching strategies, which were defined as the knowledge the teacher had to provide to their students, towards a new methodology, in which the focus has to point to the student learning process. In this way, the role of the student changes from a listener to an active agent, which acquires new skills by working not only theoretical contents but also participating in the resolution on practical situations which will definitely help to interiorize the new concepts and get a more significant knowledge.

In this process, the greater participation of practical activities in the whole TLS is obvious. Practical activities have proved to be an effective tool to provide a real knowledge not only focused on the learning of concepts, but specially focused on the learning of the "know-how" and thus allow the student to get confidence on the resolution of problems.

It is increasingly common to use terms such as competency based learning. According to U.S. Department of Education a competency is "a combination of skills, abilities, and knowledge needed to perform a specific task" [2]. Students should be able to apply their knowledge and skills to the professional and personal context [3,4]. The interest in competencies and measuring specific learning is accelerating worldwide [5]. There are different innovative activities in order to develop these competencies. These include project-based learning [6], problem-based learning [7], case study [8], etc.

One of the main changes in the implementation of the European Higher Education Area is the way to measure the teaching load. Formerly, it was referred to the period in contact between student and teacher. Nowadays a description of the qualifications is recommended in terms of the student workload [9]. Student workload can be defined as the number of working hours, which includes attending seminars, lectures, of tutorials (contact hours) and private study, preparation of projects and exams, etc. The workload associated with a credit depends on the country. In the EHEA, it is suggested that 1 European Credit Transfer System credit (ECTS credit) represents approximately 25-30 student working hours [10]. In the University of Extremadura, 1 ECTS credit corresponds to 25 student hours. In this context, activities should be carefully planned.

In this chapter we introduce the problem-based learning concept, we describe its main characteristics and implementation mode; moreover, a revision of the state of art is made, focused on the University of Extremadura, and then an statistical analysis of the current bibliographic resources available in Scopus is made. In this way, we aim to provide a frame to contextualize the book titled "Project Based Learning on Engineering: Foundations, Applications and Challenges".

2. PBL ORIGIN AND CHARACTERISTICS

Problem based learning (PBL) is a learner focused educational approach where the student extends previous knowledge to new problems through self-directed reflection, research and practice in solving a problem. According to Savery [11], for PBL to be effective, the selection of the problem is critical to the success of the approach, as well as the guidance provided by a knowledgeable mentor. Students learn through problem solving where there is usually no single right answer.

The main characteristics of PBL approach are described next [12]:

1. Students must have the responsibility for their own learning.
2. The problem used in PBL must be ill-structured and allow for free inquiry.
3. Learning should be integrated from a wide range of disciplines.
4. Collaboration is essential.
5. What students learn during their self-directed learning must be applied back to the problem with reanalysis and resolution.
6. A closing analysis of what has been learned from work with the problem and a discussion of what concepts and principles have been learned are essential.
7. Self and peer assessment should be carried out at the completion of each problem and at the end of every curricular unit.
8. The activities carried out in PBL must be those valued in the real world.

Figure 1 shows the main differences between traditional teaching model and PBL based approach.

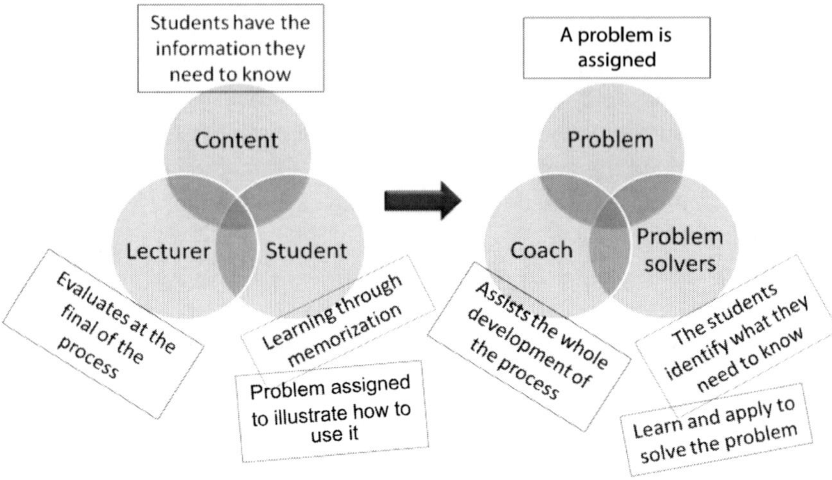

Figure 1. Characteristics of PBL process as compared to traditional learning.

3. Implementation of Problem Based Learning Procedures

The PBL model has been implemented in various ways since it was first developed, more than 30 years ago. Nonetheless, the original process described by Barrows and Tamblin (1980) [13] remains the most commonly used. According to this model, learners meet weekly or biweekly in groups of five to ten with a tutor present. Learners work through problems in a systematic way to accomplish learning objectives they have identified, and each problem occupies learners for a minimum of two sessions. Evaluation of the learners outcomes is a central feature of the process, and includes assessment of both individual accomplishment and the group´s ability to work effectively.

The whole process includes several steps: from five (as Barrows traditional approach) up to nine [14], depending on the approach considered. Independently on the model, there is consensus that in response to a presented problem learners should take actions following a scheme similar to that defined below, and teacher-coaches should modify the steps accordingly to suit the task and the students:

1. The teacher-coach sets the stage for students with real-life samples of the projects they will be doing. The scenario is defined and the gap is presented. In addition, each problem includes relevant data along with a tutor guide listing intended learning issues (learning package).
2. Students take on the role of project designers, possibly establishing a forum for display or competition. Between this stage and the following one brain storming process are useful to identify different pathways to solve the problem.
3. Students discuss and accumulate the background information needed for their designs. The group begins by identifying terms and concepts that are unfamiliar to some or all members. Terms that are unfamiliar to some students may be known to others, and so the process information sharing begins at this stage. Alternatively, the students may have resources with them to consult for quick definitions.
4. The teacher-coach and students negotiate the criteria for evaluating the projects.
5. Students accumulate the materials necessary for the project.
6. Students create their projects.
7. Students prepare to present their projects.
8. Students present their projects. (Stages 7 and 8 may appear joined in other models)
9. Students reflect on the process and evaluate the projects based on the criteria established in Step 4.

Figure 2 shows an example of multistep PBL process. As previously stated, these steps are variable depending on the author and can be adapted for each particular case.

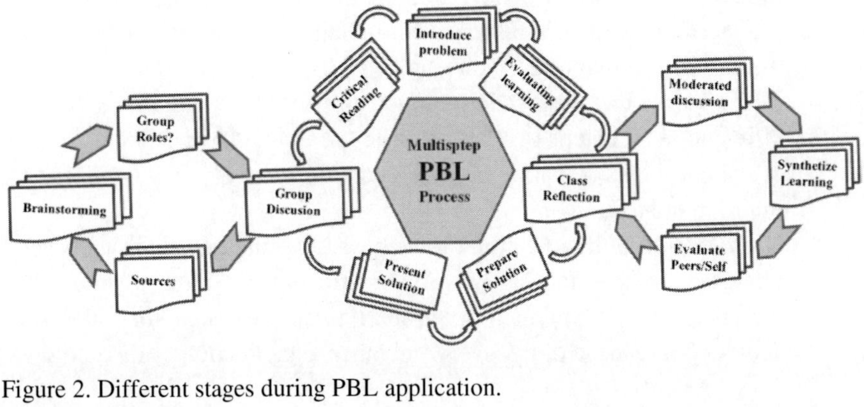

Figure 2. Different stages during PBL application.

The teacher's role is to serve as coach, guiding students to use a variety of resources, employing a strategy that is fun and motivating. The following questions are examples of those which a PBL coach should make himself:

- Do the students have a clear understanding of the task?
- Does each student have ownership of his role within the group?
- Are the students attentive and working together cooperatively?
- Are there sources that students use geared to their comprehensive level of understanding?
- Are any groups stumbling in a way that is blocking their work due to heightened emotions?

Regarding the coach's role, if the teacher is an expert in the matter in areas of the knowledge required, the group may ask for information. However, if the tutor provides information, it should be only provided in the form of isolated terms of knowledge, (since the tutor´s role is not to be transmitter of information but instead the facilitator of student-directed learning).

Many factors are needed in order to achieve a successful PBL that leads to achievement of acquisition of retrieval knowledge. Those factors are:

1) Prior knowledge: according to constructivism theory of learning, the amount prior of knowledge and its activation on time in the discussion is essential.
2) Quality of the problem: The well constructed problem will trigger and stimulate discussion.
3) Tutor behavior: The tutor skill to maintain the discussion of the problem will increase the quality of the problem.
4) Student centeredness: Motivation of the students is increased when the learning is his/her responsibility. This will lead to application of the self determination for learning.
5) Team work: the well functioning team will lead collaboratively to efficiency in all steps of PBL process. Group dynamic will ensure that every one is participating efficiently and effectively, this will lead to improve the team work.
6) Ground Rules: Organization on the basis learning ground rules is very important in maintaining the organization of the tutorial group and maintain group dynamic. By ground rules we mean the rules that should be respected by every member e.g. keeping time, how to

participate in the discussion, how the work in data collection about the learning objectives will be distributed, etc.

4. EVALUATING THE PROCESS

Assessment is a crucial part of the learning process; it reports the extent of accomplishment of the learning objectives. This should be the major drive when planning student assessment in PBL.

Evaluation of PBL process may be regarded from two points of view: the first one is related to the knowledge content or subject matter, and this can be assessed using the ordinary assessment methods used for the knowledge or cognitive domain.

The second type which is related to the process itself, the acquisition of problem solving skills, leadership, communication skills, etc. for this type, special way of assessment should be placed. This can be done after tutorial sessions, formally of informally, or at the end of the module; using check list is the common method used for collection of data for this assessment. The students should reflex about how they enjoyed about working in small groups, and how one student's idea would spawn another student's idea. They should discuss what they liked about the materials and what they found to be frustrating. They share their reflections to note what they had in common and what was special to each group or to each individual personally. They reviewed the criteria of assessment and discussed how well they met them.

5. APPROPRIATENESS OF PBL IN ENGINEERING EDUCATION

During the last decade there has been an increasing interest among the teaching community about including PBL and other teaching/learning strategies such as engineering disciplines. However, since the model was created in the Medical field, it is important to be aware of some important issues. In engineering field, the effectiveness of this model has in general been proven [16-23]. Indeed, some teachers like Perrenet [15] state that while PBL is effective in medical education because of its "encyclopedic structure," it may be more challenging in engineering because of its hierarchical structure. In medicine, the order in which various concepts are learned generally is nor critical. In contrast, mathematics, physics, and in much engineering, the order in which the knowledge is gained is critical. For example, a solid foundation on thermodynamics must be achieved before moving to design an engine. Perrenet suggest that because of this hierarchical structure, a fully integrated engineering PBL program might not produce the "right" knowledge at the right time. Letting the students all the responsibility of ordering the concepts they learn as they go, might be somehow confusing or led to bad results.

At this point it is interesting to emphasize the concept of self-directed learning, as defined by Barrows, in which it is stated that the students decide what they need to learn. In this sense the labor of the tutor has to be carefully addressed in order to ensure the correct settle of the learning objectives.

Also, while in medicine there is usually only one correct diagnosis which can be made relatively quickly, in engineering, an authentic problem could extend over long periods of time with multiple solutions.

6. PBL in Engineering Education in University of Extremadura

The Telematics group in Engineering's area of the University of Extremadura has implemented the Problem Based Learning (PBL) methodology in some of the subjects of the 4 courses degrees, which can be considered a vertical teaching methodology. They have developed an innovative teaching project designed in two phases: the first phase was implemented during the 2012/2013 academic year and the second phase during 2013/2014. In the first phase each professor implemented in their subject an innovation related to Information Technology and the second phase, all subprojects were linked as a puzzle, so that a large project was obtained as a result of the joint work of several subjects.

The main objective of this teaching project was to strengthen skills acquisition related to Information Technology and Communications through an active learning methodology as PBL. In developing each subproject, the student had to perform various tasks to acquire skills such as developing critical thinking, being able to make decisions and team work, solve problems, plan their work and have extensive specific knowledge of Information Technology and Communications.

The aim was to get a catalog of subprojects described through the PBL methodology to be used by any team of professors and develop innovative issues related to this field of knowledge.

It was developed in four phases:

Initial phase: a presentation of the problem/project this was made by relating to real life situations, as close to the professional activity. The professor had a prominent role to achieve motivation to convey to their students. Here lies part of the success of the project.

Second phase of research: each student had to seek information to perform the task.

Third phase: Implementation of self-learning and analysis.

Final stage evaluation: This stage defined a priori the minimum benefits that should be achieved to pass the assessment of the subject. The students had to play an important role during this process. In this sense some techniques were used to increase student participation, for example, peer-evaluation of each solution or the possibility of giving a global mark for a whole group that the students then had to decide how to distribute among themselves. This allowed students to test, compare and evaluate the results of individual projects before the joint evaluation with the professor.

One of the subprojects carried out by this group of professors was "Quality of Service (QoS) and Traffic Engineering (TE) in switched high-speed networks" (QUATRAFFIC) specifically in the subject Communication Services Advanced 4th year of the Degree in Telematics that treated aspects of terminal services and value-added digital hierarchies and SDH networks, broadband concepts and their application to IP networks, as well as aspects of Quality of Service (QoS) [16].

The teaching-learning activity was conducted in various work settings: whole class, group without tutor, tutor group and individual work, allowing the strength of the methodology, through an effective workout done early in the whole class and through clear instructions, groups of students worked intensively without professor intervention and outside of class time. The methodology was developed in the four phases mentioned above. However, before the activity began, the PBL was briefly described to all students, in order to inform them about its usefulness and how it can help to provide professional competences an engineer should have.

From the beginning of the activity, they were treated as members of a Networking/Communications Department to which they address a problem related to professional work.

The specific objectives to be achieved were shown to students: implement "green solutions" in order to develop their ethical responsibilities as an engineer, capacity for autonomous learning, face a real problem and be consistent with the decisions to be taken are starting a project on technology not studied, participate in a team with similar experts, ability to prepare technical meetings (brief), technically defending a solution or proposal, discussing, listen to other opinions and accept the consensus of the team and ultimately leadership groups.

The first minutes of the session, in the classroom, the whole group discussed the outstanding issues, took planning decisions for the rest of the session and reviewed the group dynamic initial. The professor only participated to give some small indication points and controlled and monitored the activity.

Apart from information and resources, the teacher-coach provided an example of "Company report" to the students through the virtual campus or other virtual fold such as Dropbox. Regarding the evaluation of the specific knowledge acquisition, it was found that the average qualification was above the mean value of other scores obtained by the same students when applying traditional teaching methods.

The evaluation of the PBL activity was performed using an anonymous questionnaire through the Virtual Campus of the subject, with descriptive questions possibilities, corresponding to the following scale: 1 = Strongly Disagree; 2 = Disagree; 3 = Undecided; 4 = Agree; 5 = Strongly Agree. The assessment in all cases has been very positive. It was found that this activity has made them more flexible in their thinking and feeling less independent work of their schoolmates.

Other subprojects developed by this group were applied in the subject "Computer Networks" (3rd year) whose contents are: the network layer in the Internet; IP Protocol and Analysis of IP packets. In any case it was aimed to get the same skills mentioned in the previous project. The working methodology, as well as materials and resources were similar to those presented above. Each group developed a software system for testing the error handling IPv4 ago. In terms of assessing the knowledge gained the following issues will be taken into account, participation in the activity, technical quality and presentation of the final report choosing to do one group member at random, classmate assessment and self-evaluation [17].

In this line the same group has proposed another teaching innovation project, in the subject of "Mobile Wireless Networks", in which three different subprojects based on the same methodology described above, were developed.

In the first project the students developed a professional project to install a WiFi network combining PBL methodology with the use of web platforms. The subproject might be capable of being verified by the relevant professional staff, so that it provided the learner's telecommunications engineer skills. In the second project they had to make a coverage map of the different WiFi networks from the center, in this case combining this methodology with a professional software. In the last project they used their mobile phones to make a series of measurements of cellular networks in an area of the city; in this they developed an innovative methodology for m-learning (learning through mobile devices) [18].

With the implementation of the three projects the students overcame most of the objectives that were determined in two ways: Evaluation of reports submitted, and presented, and test of knowledge. The results were very satisfactory, since over 90% of the students demonstrated to have acquired more than 50% of the learning objectives.

Specifically, for the subject "Standardization and Telecommunication Projects" (4th year degree of Telematics Engineering with 6 ECTS credits), with a workload of 150 hours, of which 90 are non-contact, and the remaining hours through ABPL methodology, the professor performed continuous monitoring of learning by the students. Students were required to design a draft CTI (Common Telecommunications Infrastructure) with the PBL methodology. Each group made a part of the project, so that the final outcome was a result of the work of each group. Likewise, they were encouraged to make searches for optimal solutions, which reduce consumption of materials to be installed, and a part of the project consisted of home automation solutions that reduce energy consumption of buildings, tending to build more sustainable buildings [19].

As in previous cases, the activity began with an explanation about what PBL is, and grouping students in groups 3 or 4. Each group delivered at the end of the course a scanned drawings (Royal Decree 346/2011) so that the CTI project was done by the whole class. Also, students had to develop performance standards for each group, which had to include the minimum number of meetings, the proceedings of a secretary of the group. This proceeding should contain a summary of the issues raised and resolved in each group meeting and disciplinary measures performed by the group when a group member did not attend the meetings, or was late, or did the part of labor imposed by the group. The rest of the methodology and criteria of evaluation are similar to those already described.

The results showed that 92% of students surrendered and continued the work during the development of the subject, developing teamwork skills and learning to make ICT projects, which can be performed in real life as engineers. The mean scores were higher than 6.5 out of 10. The course was not passed by 15% of students. Finally, satisfaction outcomes were measured through surveys that students held on various aspects of the course.

Another of the works presented by this group is the result of teaching innovation project "Implementation of Problem Based Learning in Engineering degree" for 2012-13 course. Specifically, it described the work done in the initiative "Connection and writing verification twisted pair network in a structured cabling system", held during the first semester of the 2012-13 in "Networking Fundamentals" class. The overall goal was that students learned to recognize structured cabling systems, and how the network infrastructure and the passive elements that compose them was organized. To achieve the objectives of the proposed learning, students were organized in groups to plan, implement and verify the installation of network cabling of a classroom demonstration of a company. In the first phase, the materials (Cat 5e/6) were provided and an initial project planning was done. During the investigation (phase 2), each group reviewed the literature provided by the professor and could make own research on methods and techniques associated with connection and verification of SCE, selecting those they deemed most useful references to help. During the project (phase 3), each working group carried out a series of practical tasks including: connected network jacks RJ-45 network in the work area and a patch panel rack 19 inches, and installation verification check using a computer network cabling. As CTI tools to support the completion of this activity, the groups had to kept a notebook computer in a collaborative way through Google Docs application and maintain a bibliographic repository of files in Google Drive. With regards to evaluation, it relied on the use of survey and rubric of Moodle 2.0 platform used in the virtual campus of the University of Extremadura modules[20].

On the other hand, in the field of teaching degree in Mechanical Engineering from the School of Industrial Engineering, particularly in the subject of "Mechanisms and Machines" this type of experience during the 2010/2011 academic year was also carried out.

Working groups of 6 students were formed. Each group was asked the statement of a PBL activity, in particular the modeling and simulation kinematics and dynamics of a mechanical system using CAE tools and verification using MATLAB results.

The procedure for pursuing the activity was similar to that described for other authors and the results were very positive for both students and for professors. According to Alonso Sánchez et al., 2011, problem-based learning is a great learning tool to take profit to the non-contact hours of student work in the context of the ECTS credit system [21].

The professors Arévalo et al. have designed the activity for the course of Materials Chemistry of the Civil Engineering Degree at the Polytechnic School. In this work, this group developed a PBL scenario as a means for the teaching-learning of a practical course on General Chemistry. Students were presented with the challenge of building a well in scarce resources situation where the quality of the materials used is not guaranteed. They had to solve the various issues involved so that the final requirements for the materials were tested and met. The analysis of the data obtained showed that the students' understood better the theoretical concepts of the course [22]. In this case, while encouraging teamwork and problem solving, laboratory work was reinforced [23].

7. The Contextual Situation of PBL in the Research Sources

The context on PBL publications in the research field is analyzed in the next section. It has to be considered that the term PBL can be referred at least with these concepts (process, problem and project based learning).

The growing of the number of documents published in the last years (Figure 3) has developed and exponential behavior, fact that indicates the increasing interest in this field. The requirement (exigency) of the editors of the different journals could be considered a quality filter, and linking these concepts, the statistical analysis of the PBL articles could be considered a method to study the evolution of this field in the academic, and research context.

The transversal nature of this field has contributed to different areas of knowledge, presenting a greater amount of research works in the applied disciplines.

The influence of PBL in the different disciplines is showed in Figure 4, presenting different distribution.

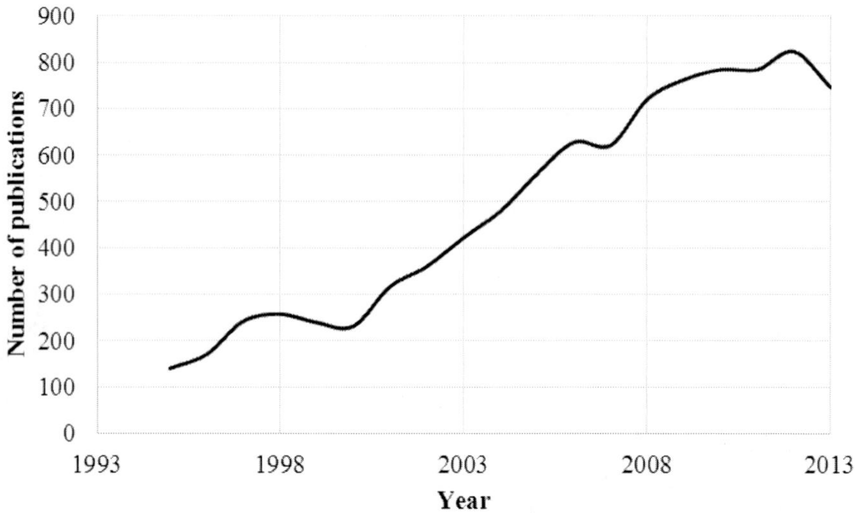

Figure 3. Project based learning, Process based learning and Problem based learning publications. Source: scopus.com.

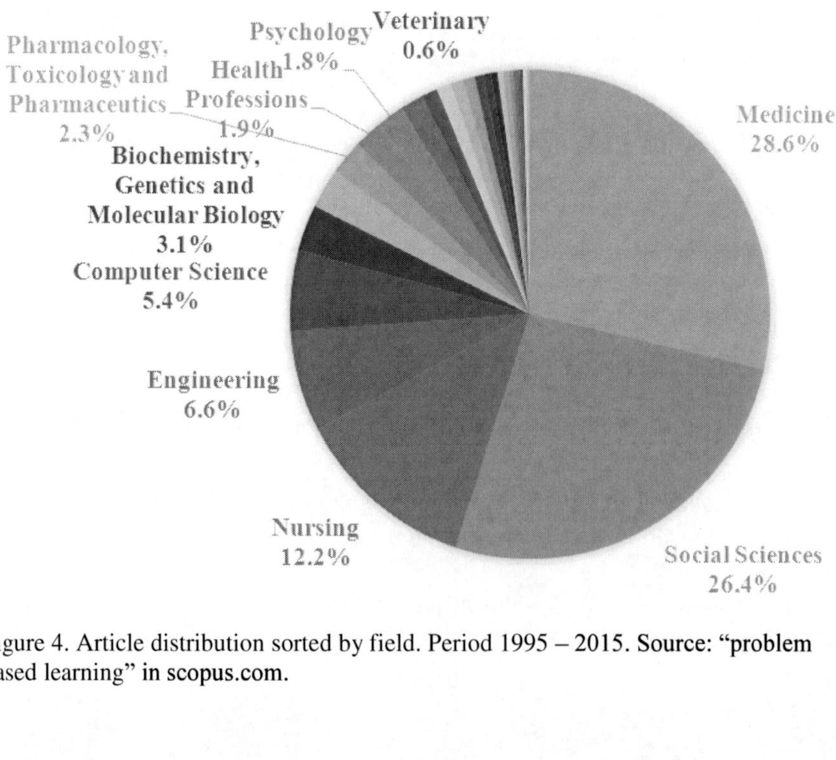

Figure 4. Article distribution sorted by field. Period 1995 – 2015. Source: "problem based learning" in scopus.com.

The major part of the publications belongs to the medical sector, which is coherent with the origin of this model, which was first applied in Nursing, if Medical, Nursing and Social Sciences are joined, their share reach more than 50% of the total, corroborating the potential and influence in the topics related directly with the human beings relationships.

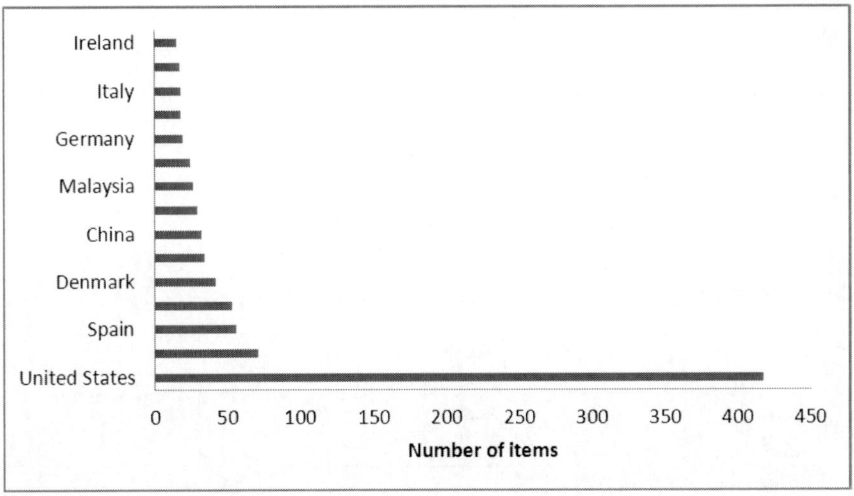

Figure 5. Distribution per countries. Period 1995 – 2015. Source: "problem based learning" in scopus.com.

The geographical distribution of PBL bibliographical sources is shown in Figure 5. In spite of the total United States dominance, focusing in the developed countries the dispersion around the continents is observed. The role of Europe countries and specifically Spain seem to be among the most advanced in terms of amount of documents published in the journals.

Not least important should be to consider the quality of the publications in the research and developed community. As this discipline is still growing in a fast mode, it is more rigorous to analyze it with resources where quality ratios are used to measure the quality level comparing with other publications related in the same field.

Some of the most relevant journals in the engineering area are shown in Table 1. A deeper analysis of the ranking of these journals shows the dispersion in terms of percentile location.

To sum up, the effects of the contextual situation about the PBL in the research context could be described considering that the fact of comparing the ratios with other publications, even in the same field, needs to be analyzed attending the methodology of the statistics.

Table 1. Relevant journals in the engineering area

Journal Name	Number of publications	5 Year Impact Factor	Education Scientific, disciplines - Category	Engineering, Electrical & Electronic - Category
International Journal of Engineering Education	71[1)]	0.434[1)]	Q4[1)]	Q4[1)]
IEEE Transactions on Education	33[1)]	1.3101[1)]	Q3[1)]	Q2[1)]
Proceedings Frontiers in Education Conference	32[2)]	0.110[2)]	-[2)]	-[2)]
European Journal of Engineering Education	24[3)]	1.040[3)]	Q1[3)]	Q1[3)]

[1)]JCR (Journal citation report).

[2)]SJR (SC Imago Journal & Country Rank), by Scopus. Non related to quartiles. One year impact factor.

[3)] SJR (SC Imago Journal & Country Rank). One year impact factor.

The structure of the journals awards the publication most cited and the indicator increases the quality directly depending of the repercussion. This seems to indicate that nowadays the PBL topics do not offer apparently a deep interest to the researchers, and in consequence the citations between each other are not increasing, nor consequently the impact factor. Nevertheless, according to the trends analyzed in the charts, it could be considered that in the next years the interest will get more and more prominence, improving the quantity and the quality of these fields.

CONCLUSION

The new educational scenario requires the inclusion of new methodologies. These methodologies should be student-centered and should be oriented to develop competencies. These characteristics make this model an ideal candidate for this new educational framework. On the other hand, while interest in PBL model is increasingly rapidly; the number of publications related to the field of engineering is still scant.

In this book experiences of using problem-based learning in engineering classes at the University of Extremadura are described and analyzed.

REFERENCES

[1] The Bologna Declaration on the European Space for Higher Education: An explanation, 1999, http://www.nvao.net/page/downloads/Bologna_Declaration.pdf, accessed 19 November 2014

[2] E. Jones, A. Voorhees and K. Paulson, Defining and Assessing Learning: Exploring Competency-Based Initiatives, U.S. Department of Education, *National Center for Education Statistics Washington*, D.C., 2001.

[3] S. Román-Suero, J. Sánchez-Martín and F. Zamora-Polo, Opportunities given by final degree dissertations inside the EHEA to enhance ethical learning in technical education, *European Journal of Engineering Education* , 38(2), 2013, pp. 149-158.

[4] F. Zamora-Polo, S. Román-Suero and J. Sánchez-Martín, From Efficiency to sustainability. Training responsible engineers in the new educational scene, *Dyna* (Spain), 85 (7), 2010, pp. 575-580.

[5] R. A. Voorhees, Competency-Based Learning Models: A Necessary Future,New Directions for Institutional Research, Willey Online Library, 2001, pp. 5-13.

[6] S.R. Gonçalves Fernandes, Preparing Graduates for Professional Practice: Findings from a Case Study of Project-based Learning (PBL), *Procedia - Social and Behavioral Sciences*, 139 (22), 2014, pp. 219-226.

[7] R. Martín-Rojas, V. J. García-Morales and M. T. Bolívar-Ramos, Influence of technological support, skills and competencies, and learning on corporate entrepreneurship in European technology firms, *Technovation,* 33 (12), 2013, pp. 417-430,

[8] G. Wilson, Constructivist learning environments: case studies in instructional design, *Englewood Cliffs, Educational Technology Publications,* New Jersey, 1996.

[9] J. R. Ruiz-Gallardo, S. Castaño, J. J. Gómez-Alday and A. Valdés, Assessing student workload in Problem Based Learning: Relationships among teaching method, student workload and achievement. *A case*

study in Natural Sciences, Teaching and Teacher Education, 27(3), 2011, pp. 619-627.

[10] European Commission, ECTS Users' guide. European credit transfer and accumulation system and the diploma supplement European Commission – Directorate General for Education and Culture, 2005, from http://europa.eu.int/comm/education/programmes/socrates/ects/guide_en.pdf. Accessed 19 November 2014.

[11] J. R. Savery. Interdisciplinary Journal of Problem-based Learning. Overview of Problem-based Learning: Definitions and Distinctions. Published online: 5-22-2006. 2006. Vol. 1 (1).

[12] "The Problem Based Learning Initiative", www.pbli.org.

[13] J. Barrel. Problem based learning. An inquiry approach.2nd Edition. Conwin Press.

[14] Stix and F. Hrbek. Teachers as Classroom Coaches: *How to Motivate Students Across the Content Areas*, ASCD Edition, 2006.

[15] J. C. Perrenet, P. A. J. Bouhuijs and J. G. M. M. Smits, The suitability of problem-based learning for engineering education: theory and practice. *Teaching in higher education*, 5(3), 2000, pp. 345-358.

[16] J. Carmona, J. C. González, M. A. Martín, P. J. Pardo, J. Arias, J. F. Valenzuela, Problem-based learning for skills development in advanced communications, 8th Iberian Conference on, Published in: Information Systems and Technologies (CISTI), *Lisboa Publisher: IEEE*, 19-22 June 2013, pp. 1–7

[17] J. Arias, J. F. Valenzuela, J. Carmona, J. C. González, M. A. Martín, P. J. Pardo, Project-based learning for Error Handling in IPv4Packets, 8th Iberian Conference on, Published in: Information Systems and Technologies (CISTI), *Lisboa Publisher: IEEE*, 19-22 June 2013, pp. 1–5

[18] P. J. Pardo, J. F. Valenzuela, J. Arias, J. Carmona, J. C. González, M. A. Martín, Proyecto de innovación docente en comunicaciones inalámbricas, Third International Congress of University Teaching, *Vigo,* 20-22 June 2013, p. 31.

[19] J. C. González, M. A. Martín, P. J. Pardo, J. F. Valenzuela, J. Arias, J. Carmona, Proyecto de innovación docente en normalización y proyectos de telecomunicación, Third International Congress of University Teaching, Vigo, 20-22 de Junio de 2013, p. 34.

[20] M. A. Martín, J. Arias, J. Carmona, J. C. González, P. J. Pardo and J. F. Valenzuela. Conexión y verificación del cableado horizontal y tomas de

red (UTP Cat 5e/6) en un Sistema de Cableado Estructurado. XXI University Conference on Educational Technology, *Valladolid*, 27-28 June 2013, pp. 59-60.

[21] F. J. Alonso, D. Rodríguez, G. Galán, C. Gragera and M. Reino, Aprendizaje basado en problemas mediante herramientas CAE en un curso de mecanismos y máquinas la evaluación de la calidad de la investigación y de la educación superior, FECIES: VIII FORO, Santander, 31 May-3 June 2011, ISBN-978-84-694-3488-8, pp: 608-609.

[22] M. J. Arévalo, G. Silvero, I. López, Design and implementation of a problem-based learning scenario for a general chemistry lab course for civil engineering undergrads, *EDULEARN-International Conference on Education and New Learning Technologies*, Barcelona (Spain), 01/07/2013, pp. 5515 – 5520.

[23] J. Martínez and G. Pinto, Reflexiones sobre los contenidos de la asignatura común de Química en los nuevos grados de distintas ingenierías, La Química como materia básica de los Grados de Ingeniería, Madrid, 2009, Publisher ETSI Industriales 978-84-7484-216-6, pp. 29-38.

In: Project Based Learning on Engineering ISBN: 978-1-63482-217-6
Editors: J. Valenzuela-Valdés et al.

Chapter 2

FOUNDATIONS ON PBL

Juan F. Valenzuela-Valdés,* Rafael M. Luque-Baena and Francisco Luna

Departamento de Ingeniería de Sistemas Informáticos y Telemáticos
Centro Universitario de Mérida, Universidad de Extremadura
Av. Santa Teresa de Jornet 38, 06800 Mérida (Spain)

ABSTRACT

This chapter introduces the main features of problem-based learning (PBL), as well as a brief revision of the related literature. First, the main facets of this methodology are outlined. Different evaluation strategies for PBL together with several particular implementations are discussed next. The final part of this chapter is devoted to presenting rubrics to assess the competences achieved with PBL.

INTRODUCTION

The very beginning of problem based learning (PBL) at the university level dates back to the mid-sixties of the last century in the Faculty of Medicine at the MacMaster University in Canada, where it was initially developed by Barrow [1-5]. In Europe, PBL was first implemented in the

* juanvalenzuela@unex.es.

University of Maastricht (Netherland) in 1974, also in Medicine, following the guidelines used in the MacMaster's experience. This European university, however, started introducing PBL in six other disciplines encouraged by the success obtained. The reader is referred to [6-9], where several comprehensive surveys of the works carried out during the first 25 years of the implementation of the methodology are presented.

Despite being 40 years old, PBL is still prevailing as it is the perfect complement to the novel learning strategies, such as cooperative learning, e-learning or m-learning (learning via mobile devices). For example, cooperative learning promotes working in groups and develops the social abilities of the members, one of the most demanded capabilities in the current labour market. The benefits of this combination are defined in [10] as "Combining PBL with cooperative learning provides a mechanism for students to maximize their own and other group members' learning by working in teams to accomplish a common task or goal". Different implementations of cooperative PBL can be found in [11-12]. Another successful integration is based on using PBL and e-learning [13-19].

FOUNDATIONS

PBL has defined a novel strategy in which the students, besides learning on a given subject, also have to acquire a set of competencies and basic skills that are known to be quite useful for their professional career and personal development. In this methodology, the students are organized in groups with few members that develop, elaborate, and address a given problem closely related to the topics of the subject or degree, in a highly autonomous way. Typical issues such as the size of the groups, the type of problems stated, how many problems are addressed by a student in a given subject, what are the items to be evaluated (and that later determines the grade), etc., depend on each particular implementation of PBL. This methodology is applicable to different branches of knowledge as engineering, business management, and legal sciences. If the traditional learning methods are based on the transmission-acquisition of knowledge, PBL aims at promoting the student to be autonomous and get along in its professional life by identifying and tackling problems, understanding the impact of its own professional activities and their ethical implications, managing data, and devising strategies. . As a result, the student has to be able to make use of this knowledge acquired in the learning process. PBL is a dynamic approach to teaching, where the students learn to explore challenging, real-world problems, while they pursue a transversal syllabus and improve their working skills in small groups. PBL is active and committed, thus promoting the students to seek for a deeper knowledge of the topics involved in the subjects. The research also points out that students are more prone to retain the acquired knowledge with PBL than with the classical teaching focused on textbooks. The creation and progress of workgroups also develop confidence and self-guiding skills.

PBL is especially well suited for highly dynamic technological domains because the students have to successfully face and get adapted to such a competitive scenarios in their professional life. But this general framework, which is usually developed under a particular teaching methodology, can be implemented in very different ways. The truly interesting point is that the base problem in PBL must be related to one o more topics of the course subject or discipline the student is signed up for, but its scenario and its contents must belong to the real world, with connections either to previous experiences or the subsequent professional activities, making it attractive, interesting, and motivating.

METHODOLOGY

The methodology introduced by PBL has evolved since its beginnings, being its classical version that had been used in the faculties of Medicine more than fifty years ago. This initial version is known as the Maastricht methodology [20], which is composed of 7 steps that take place both in classrooms and libraries. There will be further details in Section 3.1. A variation of the Maastricht methodology has been developed in several Spanish universities, but using just 4 phases with rather small differences among them. Section 3.2 includes a description of the implementation in the *Centro Universitario de Mérida* at the University of Extremadura, which follows this mentioned variation.

We want to remark here that PBL is steadily moving on and has undergone a continuous and fast evolution mainly because of the revolution of the information and communication technologies (ICTs). Nowadays, ICTs have provided students with countless sources of information, not always reliable and confident. Therefore, they have to be endowed with the ability of getting along in such rich-information environments, what has led to a competency that our students must train during their university period so as to be prepared as competent and modern professionals. Also, the new tools for online discussion, the collaborative work, and the integration of information from different authors have provided us with powerful educational tools, allowing us to deploy PBL activities in virtual environments where the students learn to perform well as active members of a virtual learning community. Finally, working in groups has moved from being in-person and developed in the classroom, to being carried out outside and, eventually, by using virtual communication tools. These methodological versions of PBL emerged in the last years will be discussed separately in different chapters of this book.

Classic Methodology of Maastricht

This classical method has seven steps that are implemented in classrooms with few students (ideally, less than 20). The methodology can be divided into three different phases: in the initial phase (five steps), the students analyze the problem to be solved, the second one is devoted to studying the tools they need for the third phase, which finally lies in addressing the problem itself. Different groups work on different phases at the same time. The 7 steps of this methodology are detailed next:

- Step 1: Read the problem stated and discuss the issues they already know and understand, as well as those they don't.
- Step 2: The problem to be addressed is defined accurately.
- Step 3: A first approach for addressing the problem is tackled. The technique usually applied is brainstorming so as to share and discuss several points of view.
- Step 4: Steps 2 and 3 are revisited and several approaches to solve the problem are stated. All the initial hypotheses or research lines arisen in the discussion have to be analyzed, evaluating pros and cons of each one.
- Step 5: A consensus with the final solution that will be finally implemented is reached. The roadmap of this solution is agreed, as well as the roles of each group member. The tutor has to make sure that the roadmap is suitable to some extent.
- Step 6: Search for information and personal study. The students look for information related to the problem stated.

- Step 7: Discussion and resolution of the problem. The students argue on the solutions found in the previous step. They have to value properly both the confidence and relevance of the different sources of information. If required, a new loop of research is performed until enough information is gathered to address the problem. Finally, all the research findings are shared among the groups.

FOUR-PHASES METHODOLOGY

The four phases of this methodology are described separately in the next subsections.

Initial Phase

The role of the lecturer is critical in this phase as the students have to be motivated properly. In order to do so, the lecturer will first present the target problem/project, connecting it to real-world scenarios, then s/he will ease the project scheduling, initially guiding the search for related information. The aim is that the students learn and embrace the background concepts as well as set out a working plan.

Research

In this second phase, each student will look for information on their own. With such information available, they will define an initial approach for the resolution of the stated problem. A constructive discussion has to take place between students and lecturer about the solution proposed and the procedure followed to end up with a successful project. The students are also responsible for scheduling the meetings with the lecturer to ask for guidance and advice.

Project Development

This is the most challenging step for the students, but also the one in which they will develop their ability for self-learning and analysis, as they will have to foresee all the issues that the project development will pose. A continuous guidance from the lecturer is required too.

Evaluation

The grade of a project will depend on its final functionality. Indeed, a minimum set of operations, defined as priori, is required to pass the subject, and the grade is increased as this functionality grows up to either advanced or excellent. The functionality can be self-evaluated so that the students are allowed to know the expected grade of their project beforehand (constant feedback). The final evaluation takes place with the lecturer in a predefined date. As stated before, the grade of the student will depend on the implemented functionality and its performance.

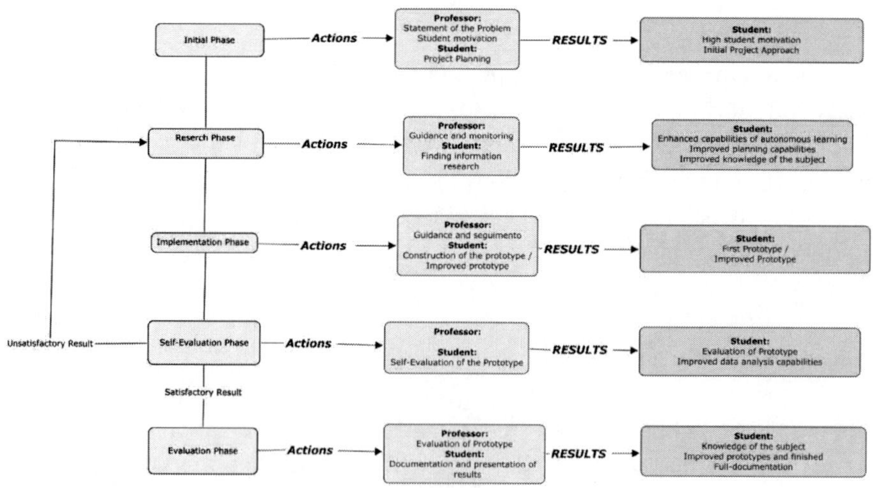

Figure 1. Methodology of the project developed.

EVALUATION

Starting from the famous statement, "You get what you assess" [21], the evaluation of PBL must be consequent with the developed methodology, i.e., the students' grades must be based on the solutions given to the problems proposed. That is, what must be evaluated are the acquired competencies. This issue has been analyzed by different schools of engineering around the world; in fact, the committee in charge of issuing the certification of engineering programs, ABET [22], has published the competencies that any engineer must acquire to obtain such certification, promoting the development of many innovative teaching projects during the last years. The work of Shuman [23] has gathered the ways of evaluating competencies worldwide, being rubrics one of the most widely used. The following tables I, II and III are devoted to presenting the three rubrics adopted to evaluate three different competencies developed in the *Centro Universitario de Mérida*. More about rubrics use and its evaluation are developed in the following chapters, thus in this chapter only introduce the rubric used.

RUBRIC – ORAL PRESENTATION

Item	Grade: A Excellent (3)	Grade: B Good (2)	Grade: C Fair (1)	Grade: D Unsatisfactory (0)
Content present-ation	Present each item properly according to its relevance, sharing out the available time, and explaining them correctly a succinctly.	Present each item, but the time has not been suitably distributed among them, and/or do not explain the items properly.	Present all the items, but eventually in disorder or with a lack of explanation for some of them. The presentation gets to explain the proposal, but vaguely and incompletely.	Present the items disordered, paying attention to just a few of them. Little-to-no information is provided on the topic and how it has been/will be performed.
Language usage	The style used is suitable for the nature of the talk. The way the contents are explained and the vocabulary used meet perfectly the context where it takes place.	In general, the style is adequate, but the speaker eventually uses simplistic ideas or vocabulary when more technical words are expected. Conversely, the vocabulary is highly technical to explain simple ideas.	Frequently, the style of the speech does not consider the context where it takes place. It often introduces inadequate ideas and/or vocabulary, given the nature of the speech.	The level and vocabulary of the speech are not adapted at all to the scenario. It is either too simple or too specialized with respect to the nature of the speech.
Quality of the present-ation	The slides include relevant and creative animations that attract the attention of the audience.	The slides used visual elements, but they are not organized properly because of either the shape or the size.	The slides uses few visual elements that with little contribution to the presentation (the figures are randomly chosen or their size is not appropriate).	The presentation is tedious, with no visual elements that attract the attention from the audience.

(Continued)

Item	Grade: A Excellent (3)	Grade: B Good (2)	Grade: C Fair (1)	Grade: D Unsatisfactory (0)
Non-verbal abilities	The speaker reinforces the message and gets the attention of the audience by using the oral presentation techniques effectively: eye contact, tone modulation, strengthen the verbal message with gestures, no pet words, etc.	The speaker uses the oral communication techniques properly, but it eventually misses eye contact with the audience, use pet words or the volume is not adequate.	The speaker does not support the speech with oral communication techniques. It is, however, able to explain the basic elements of the talk, but it does not face the audience, does not modulate the voice, does not use the adequate volume, or use pet words.	The speaker is so nervous, reads the note pages of the slides and does not use any oral communication technique. It is actually difficult for the audience to pay attention.

Rubric – Writing Assessment

Item	Grade: A Excellent (3)	Grade: B Good (2)	Grade: C Fair (1)	Grade: D Unsatisfactory (0)
Objectives	The manuscript details each of the proposed objectives in clear and concise way. The discussion is unambiguous and prioritizes properly the sequencing so. As a consequence, it is easy to check whether the objectives have been achieved.	The manuscript enumerates and describes the objectives properly, but the clarity and concision of some of them hinder the reader from check easily whether they have been achieved.	The objectives of the work are presented in an excessively general way and/or they are not sequenced properly.	The manuscript explains the objectives vaguely, so it is hardly evaluable whether they have been achieved in the work.
Methodology	The manuscript fully describes and argues the methodology used. The detailed procedures fit for the stated objectives and allow them to be achieved.	The manuscript briefly describes the methodology used, but with coherent arguments. The detailed procedures fit for the stated objectives and allow them to be achieved.	The manuscript describes the methodology in a highly simple way and with a clear lack of arguments. The procedures are appropriate for the stated problem, but they do fit for all the proposed objectives .	The manuscript scarcely describes the procedures, using an unsuitable writing. The main/all objectives cannot be achieved.

Item	Grade: A Excellent (3)	Grade: B Good (2)	Grade: C Fair (1)	Grade: D Unsatisfactory (0)
Presentation of the results	The manuscript excellently explains the results, which are correct and come naturally out from the used procedure. The presentation is clear and concise, using the most recent tools and including visual elements that improve the understanding.	The manuscript uses sophisticated tools for presenting the results, which fits for every context, such as using graphics that clarify and increase the information. The results are correct and complete, and are described using the appropriate notation (and units).	The manuscript uses simple tools, such as tables, diagrams, and basic graphics for presenting the results. These results are correct and complete, but they include numerical or notation errors.	The manuscript does not include any results or they are either partially or fully erroneous, with major numerical or notation flaws. The results are presented in a slightly effective way.
Discussion of the results	The manuscript explains and justifies the obtained results, and establishes connections with previous knowledge and similar reality from a wider context.	The manuscript explains the obtained results, connecting them to the knowledge that justifies then within the considered reality and in analogous scenarios in the same context.	The manuscript explains the obtained results by connecting them only within the considered reality.	The manuscript does not provide a coherent interpretation of the results or connect them to the knowledge can be used for their justification.
Quality of the presentation	The manuscript does have an index and numbers the pages properly. It is properly formatted (according to the instructions provided), includes tables and readable figures, and it is well organized.	The manuscript includes an index and numbers the pages properly. The formatting and organization have several errors, but they do not make the reader to get distracted.	The manuscript has a low quality that does not meet the requirements provided. The errors included hardly allow reading the work.	The manuscript has a very low quality, plenty of typographic errors and with a bad formatting that clearly difficult the reading and understanding or the work.
Writing	The student writes clearly, composing syntactically corrected sentences, without typographic errors and using punctuation properly.	The student writes correctly in general, but eventually makes small typographic errors, uses punctuation improperly, or writes down syntactically uncorrected sentences.	The student does not write clearly, with several syntactically uncorrected sentences, making several typographic errors, and not using proper punctuation often.	The student writes in disorder, and the plain text includes many syntactic, typographic, and punctuation errors.

(Continued)

Item	Grade: A Excellent (3)	Grade: B Good (2)	Grade: C Fair (1)	Grade: D Unsatisfactory (0)
Language usage	The student uses an appropriate style for the type of document. The way the contents are explained and the vocabulary used are perfectly adapted to the context.	In general, the style used is appropriate, but it eventually introduces ideas and vocabulary rather simple when more technical terms are expected; on the other hand, the language is rather technical to explain basic concepts.	The style used is frequently out of the context in which it takes place. The ideas and vocabulary used is often not appropriate for the nature of the communication.	The level and vocabulary are not adapted at all to the context. It is either too simple or too specialized and technical for the nature of the communication.
Biblio-graphy	The bibliographic references for all the related papers are all provided in the manuscript, using the correct formatting.	The bibliographic references for all the related papers are all provided in the manuscript, but some of them are not properly formatted.	The manuscript cites the used references but, even they are complete, there is missing or incorrect bibliographic information.	The manuscript does not include any bibliographic reference of, at least, a subset of the used cites.

EVALUATION RUBRIC – WORKGROUP

Item	Grade: A Excellent (3)	Grade: B Good (2)	Grade: C Fair (1)	Grade: D Unsatisfactory (0)
Prepara-tion	The student brings all the required material to the meetings and is always ready for start working.	The student almost always brings all the required material to the meetings and is almost always ready for start working.	The student almost always brings the required material, but it takes some time for him/her to start working.	The student often forgets the material or is not ready for start working.
Contri-butions	The student does always provide the group with useful ideas. S/he is a leader that contributes with a large effort.	In general, the student provides the group with useful ideas when s/he gets to participate. S/he is a strong member of the group when makes an effort.	The student provides the group with useful ideas eventually. S/he is a satisfactory member that does what is asked for.	The student seldom provides the group with useful ideas. S/he can deny participating.
Shared work	The student always listens, shares and supports the effort of others. S/he tries to keep the members working jointly for the group.	The student usually listens, shares, and supports the effort of others. S/he is not a problem for the group.	The student sometimes listens, shares, and supports the effort of others, but s/he eventually is not a good member of the group.	The student seldom listens, shares, and supports the effort of others. Often, s/he is not a good member of the group.

Item	Grade: A Excellent (3)	Grade: B Good (2)	Grade: C Fair (1)	Grade: D Unsatisfactory (0)
Attitude	The student never criticizes publicly the project or work of others. S/he always has a positive attitude to work.	The student seldom criticizes publicly the project or work of others. S/he usually has a positive attitude to work.	The student occasionally criticizes publicly the project or work of other group members. S/he has a positive attitude to work.	The student often criticizes the project or work of other group members. S/he seldom has a positive attitude to work.
Problem resolution	The student seeks and suggests solutions for the problems.	The student polishes the solutions proposed by others.	The student neither suggests nor polishes any solution, but s/he is willing to deal with solutions proposed by others.	The student neither addresses nor helps others to solve problems. S/he just leaves other to do his/her tasks.
Focus	The student keeps focused on the scheduled work. S/he is quite autonomous and self-directed.	The student keeps focused on the work most of the time. Other group members can rely on this student.	The student keeps focused on the work sometimes. Other group members must reprimand, push, and remind the student to keep focused.	The student seldom keeps focused on the work. S/he leaves others to do his/her tasks.
Time usage	The student uses the time properly to ensure that his/her tasks are performed on time. The group does no have to reschedule tasks, extend the deadline, or work on his/her tasks because of the delay.	The student uses the time properly during the entire project, but s/he has incurred in some delay. The group does no have to reschedule tasks, extend the deadline, or work on his/her tasks because of the delay.	The student is usually late, but his/her tasks are always completed on time. The group does no have to reschedule tasks, extend the deadline, or work on his/her tasks because of the delay.	The student seldom meets the deadline of his/her tasks, and the group has to extend the deadline or work on his/her tasks because the misuse of the time.
Quality of the work	The student provides a high quality work.	The student provides a quality work.	The student provides work that, occasionally, requires revision, or has to be reworked by other group members to ensure its quality.	The student provides a work that, in general, requires revision or has to be reworked by other group members to ensure its quality.
Inter-depend-ence	Cooperative negotiation style. The student has a clear interest in working together to reach a common objective.	The student has difficulties to cooperate, but s/he knows that working together is required.	Independent thinking. The student does not contribute actively to the group and it has a negative impact on the project.	Highly competitive negotiation style. The student is very independent and contributes nothing to the joint work.

(Continued)

Item	Grade: A Excellent (3)	Grade: B Good (2)	Grade: C Fair (1)	Grade: D Unsatisfactory (0)
Group management	The student, as a leader, has developed the interaction and cooperation among group members to reach a well-performing group.	The student, as a leader, made a large effort to develop the team, and it worked as expected.	The student, as a leader, made a slight effort to develop the team, but it did not work as expected.	The student, as a leader, did not develop the team, and the group did not work.
Reflection and self-evaluation	The student thinks about his/her work, shares it, and draws conclusions on the way to improve his/her work in the team.	The student analyzes periodically his/her work, shares it, but there are no clear benefits based on the adopted corrective actions.	The student thinks about his/her work, share it with other, but s/he does not propose any corrective action.	The student never thinks about his/her work in the group, or does not share it with others.
Leadership	The student shows natural leadership abilities far beyond the expectations, taking the initiative and guiding others. S/he adopts a main role at establishing the goals of the group and the project scheduling.	The student applies a good leadership and can guide others.	The student assumes the leadership with a very limited ability, and requires orientation.	The student does not show any leadership ability. S/he leaves other to establish the objectives and the project scheduling.

CONCLUSION

In this chapter, we have first introduced PBL, revisiting its original developments in higher education, and then its foundations have been presented. The next part has been devoted to describing a classic methodology and the particular methodology of PBL used in *Centro Universitario de Mérida* at Universidad de Extremadura. We have paid a lot of attention to the evaluation of the projects, as it is the most important part of any activity for the students. The goal of PBL is to acquire skills, so skills must be evaluated. Among the existing options for evaluating skills, rubrics are one of the most widely used. In this chapter, thee rubrics for evaluating, respectively oral presentation skills, writing skills and workgroup skills are introduced.

REFERENCES

[1] Barrows HS, Tamblyn, R. *Problem-Based Learning: An Approach to Medical Education.* New York: Springer, 1980.

[2] Barrows HS. *How to Design a Problem-based Curriculum for the Preclinical Years.* New York: Springer, 1985.

[3] Barrows HS. *The Tutorial Process.* Springfield, IL: Southern Illinois University School of Medicine, 1988.

[4] Barrows HS. *Simulated Patients.* Springfield, IL: C.C. Thomas, 1971.

[5] Barrows HS. An overview of the uses of standardized patients for teaching and evaluating clinical skills. *Academic Medicine* (1993); 68(6):399-405.

[6] Barrows, H. S. (1996), Problem-based learning in medicine and beyond: A brief overview. New Directions for Teaching and Learning, Volume 1996, Issue 68, pages 3–12, Winter 1996. doi: 10.1002/tl.37219966804.

[7] Albanese MA, Mitchell S. Problem-based learning: a review of literature on its outcomes and implementation issues. *Acad. Med.* 1993 Jan;68(1):52-81.

[8] Norman, G.R., and H.G. Schmidt, "The Psychological Basis of Problem-Based Learning: A Review of the Evidence," *Academic Medicine*, Vol. 67, No. 9,1992, pp. 557-565.

[9] Thomas, J.W., A Review of Research on Project-Based Learning, San Rafael, CA: Autodesk Foundation, 2000.

[10] Mohd-Yusof, K.; Hassan, S.AH.S.; Jamaludin, M.-Z.; Harun, N.-F., "Cooperative Problem-Based Learning (CPBL): A practical PBL model for engineering courses," Global Engineering Education Conference (EDUCON), 2011 IEEE , vol., no., pp.366,373, 4-6 April 2011.

[11] Sallaberry, C.; Nodenot, T.; Laforcade, P.; Marquesuzaa, C., "Model Driven Development of Cooperative Problem-Based Learning Situations Implementing Tools for Teachers and Learners from Pedagogical Models," System Sciences, 2005. HICSS '05. Proceedings of the 38th Annual Hawaii International Conference on, vol., no., pp.40c,40c, 03-06 Jan. 2005.

[12] Maskell, D.L.; Grabau, P.J., "A multidisciplinary cooperative problem-based learning approach to embedded systems design," Education, IEEE Transactions on, vol.41, no.2, pp.101,103, May 1998.

[13] Miao, Y.; Haake, J.M., "Supporting problem based learning by a collaborative virtual environment: a cooperative hypermedia approach,"

System Sciences, 2001. Proceedings of the 34th Annual Hawaii International Conference on , vol., no., pp.10 pp.,, 3-6 Jan. 2001.

[14] Quesada, J.; Calvo, I; Sancho, J.; Sainz, J.A; Sanchez, J.; Gil-Garcia, J.M.; Sebastian, R.; Castro, M., "Combining moodle and redmine as e-learning tools in Project Based Learning of Industrial Electronics," e-Learning in Industrial Electronics (ICELIE), 2013 *7th IEEE International Conference on ,* vol., no., pp.86,91, 10-13 Nov. 2013.

[15] Ching-Pu Chuang; Chih-Ming Chen; Chin-Ming Hong; Yu-Ju Lin, "Effective problem-based learning supported by digital library," *Electrical and Control Engineering (ICECE),* 2011 International Conference on , vol., no., pp.6542,6546, 16-18 Sept. 2011.

[16] Yajie Wang; Hongyu Du; Ying Hao, "A case study of Problem-Based Learning instruction design under web-based environment," Education Technology and Computer (ICETC), 2010 2nd International Conference on , vol.1, no., pp.V1-586,V1-591, 22-24 June 2010.

[17] Ali, Z.; Samaka, M., "ePBL: Design and implementation of a Problem-based Learning environment," Global Engineering Education Conference (EDUCON), 2013 *IEEE,* vol., no., pp.1209,1216, 13-15 March 2013.

[18] Secundo, G.; Elia, G.; Taurino, C., "An e-Learning System Supporting the Problem-Based-Learning Approach: the Case of "Virtual eBMS"," Database and Expert Systems Applications, 2007. DEXA '07. 18th International Workshop on, vol., no., pp.668,672, 3-7 Sept. 2007.

[19] Garcia-Robles, R.; Diaz-del-Rio, F.; Vicente-Diaz, S.; Linares-Barranco, A, "An eLearning Standard Approach for Supporting PBL in Computer Engineering," Education, IEEE Transactions on, vol.52, no.3, pp.328,339, Aug. 2009.

[20] Schmidt, H.G. (1983). Problem-based learning: rationale and description. *Medical Education*, 17, 11-16.

[21] Resnick, L. and Resnick, D. (1989). Assessing the Thinking Curriculum: New Tools for Educational Reform. Washington, D.C.: National Commission on Testing and Public Policy.

[22] ABET, Criteria for Accrediting Engineering Programs, Engineering Accreditation Commission available online on: http://www.abet.org/uploadedFiles/Accreditation/Accreditation_Process/Accreditation_Documents/Current/eac-criteria-2012-2013.pdf.

[23] Shuman, L., Besterfield-Sacre, M., and McGourty, J. "The ABET 'Professional Skills' – Can They Be Taught? Can They Be Assessed?," *Journal of Engineering Education,* January 2005, pp. 41-55.

In: Project Based Learning on Engineering ISBN: 978-1-63482-217-6
Editors: J. Valenzuela-Valdés et al.

Chapter 3

PROBLEM DESIGN, IMPLEMENTATION AND RESULTS OF PBL IN TELEMATICS ENGINEERING

Javier Carmona-Murillo*, Miguel A. Martín-Tardío and Juan Arias-Masa

Departamento de Ingeniería de Sistemas Informáticos
y Telemáticos Centro Universitario de Mérida,
Universidad de Extremadura. Mérida (Spain)

ABSTRACT

The Problem Based Learning methodology is increasing its popularity day by day. Nowadays, a lot of Universities implement this methodology to develop the skills of their students. Moreover, it is known that in PBL activities, the quality of the problem is a key challenge to achieve the learning objectives. Both the design and interest of the problem influence the motivation of the students. Similarly, the development and implementation of the problems are also important for the PBL activity to work well. Over the last few years, newly structured Problem Based Learning methodology has been implemented in the Telematics Engineering degree at the University of Extremadura, Spain. In this chapter, the experience of this PBL model is described through three case studies. The aim of the chapter is to detail three case studies of

* Emails: jcarmur@unex.es, matardio@unex.es, juanaria@unex.es

problem design and implementation of PBL in the context of these experiences. In order to quantify our experiences, some results are given in terms of student validation and academic achievement.

1. INTRODUCTION

The National Agency for Quality Assessment and Accreditation (ANECA) was established to guarantee the quality of education. As a result, many varied programs and projects have been undertaken by different universities and have started to show results. The PBL is not a new idea, Barrows, is a pioneer in medical projects [1-4]. In Spain the pioneer in this type of innovation in engineering is Miguel Valero in Castelldefels School [5-7]. The PBL encourages independent learning and collaborative learning that allows students to acquire skills.

At the University of Extremadura project-based problems or learning (PBL) is one of the most dynamic methodologies currently employed. Several of the projects undertaken at the Centro Universitario de Mérida highlight this (CMU) [7]. This chapter will describe three experiments developed during 2012-13 and 2013-14 courses for the Degree in Telematics Engineering within the following subjects:

- Networking Fundamentals: Introduction to computer networks. Basic principles of network architecture. Introduction to OSI model. TCP / IP protocol families.
- Computer Networks: Network layer. Practical applications of network services. Services oriented and connectionless. Transport Layer. And theoretical knowledge and actual programming of the Transport layer. TCP and UDP protocols.
- Advanced Communication Services: Terminal Services and value added. SDH digital hierarchies and networks. Defining broadband concepts. Applications with IP networks. Quality of integrated / differentiated services.

These subjects belong to different classes. Furthermore it is necessary that students enrolled in these courses acquire key skills that the University of Extremadura requires for all degrees:

- CT1: Planning capability and organization of staff. The ability to work in situations with a lack of information and/or with time constraints and/or resources.
- CT2: Communicate effectively adapted to the socio-economic context, both in writing and orally in their own language, knowledge, procedures, and results with special emphasis on writing technical documentation.
- CT3: Ability to communicate effectively in English.
- CT4: Ability to make decisions based on objective criteria (available experimental data or simulation).
- CT5: Ability to analyze, critique, synthesis, evaluation and problem solving.
- CT6: Capacity interpersonal relationship.
- CT7: Ability to find, relate and structure information from various sources and integrate ideas and knowledge.
- CT8: Ability to adapt to new situations and/or flexibility in organizational or technological changes.
- CT9: Leadership skills, ability to influence and motivate others effectively using available resources.
- CT10: Take initiative to provide and/or evaluate alternative or novel solutions to problems, demonstrating flexibility and professionalism when considering different evaluation criteria.
- CT11: Motivated by the quality and continuous improvement and act strictly in the professional development.
- CT12: Negotiation skills, ability to convince and accept other views.
- CT13: Motivated by professional achievement and to meet new challenges as well as a broader view of the possibilities of a career in the field of Telematics Engineering.
- CT14: Acting on professional development with responsibility and professional ethics and in accordance with current legislation.
- CT15: Ability to integrate quickly and work efficiently in single-discipline and multidisciplinary teams, assuming different roles and responsibilities with full respect for fundamental rights and equality between men and women
- CT16: Ability to undertake further study with a high degree of autonomy.

Finally, note that most of the teachers of Telematics Engineering were involved in a teaching innovation project called "Learning Implementation Issues based on the Degree in Telematics." This project has been being developed throughout 2012-13, 2013-14 and 2014-15; whose main objective is to enhance the acquisition of skills through active learning methodologies such as PBL.

The chapter starts with a related work which shows the application of PBL methodology. After the objectives have been developed, generic skills and specific competencies within these PBL activities are explained. Then three case studies of the application of PBL methodology are described for each of the subjects that have been mentioned above. Finally, the chapter concludes with a discussion of the results and conclusions of the authors, based on the feedback obtained through student evaluation of the suitability of this type of activity in this context.

2. Related Works

The application of PBL methodology goes back over 30 years of development, mainly in teaching related to health sciences [8]. However, its application in the field of engineering does not begin until a decade later [9-12]. The evolution of the implementation of PBL methodology between these two fields or areas of knowledge, particularly in the area of North America, as reflected in the work [13] is very interesting. Elsewhere [14], the authors review the application of this methodology at the international level in order to introduce it into the study of engineering in Malaysia. The versatility of this methodology for engineering education [15-17] can be seen in the literature with works in several engineering fields such as aerospace, electrical and chemical engineering, as well as computer science or telecommunications engineering [18-24].

3. Aims and Skills

The principles for quality teaching shown below are sorted by relevance according to the authors in this work [25]:

1. Encourages contact between students and teachers.
2. Promote cooperation among students.
3. Encourages active learning.
4. Provide "feedback" on time.
5. Dedicate time to more important tasks.
6. Projects to students high expectations.
7. Respects diverse talents and ways of learning.

Although these principles are somewhat ambiguous and difficult to measure, they can relate well with the skills that this work has been trying to develop:

1. Ability to learn independently.
2. Ability to teamwork.
3. Facing a real problem and being consistent with the decisions taken.
4. Start a project on technology as yet unstudied.
5. Defend technically a solution or proposal.
6. Find a solution within the means available.
7. Participate in a team with similar experts.
8. Discuss providing arguments, learning to listen to other opinions and accepting the consensus of the team.
9. Finding and selecting appropriate information to solve the problem.
10. Ability to prepare a written plan to respond to the proposed project.
11. Ability to present a work.
12. Ability to allocate tasks and manage individual and group responsibility.

Even though the first principle appears not to be included in these skills, and neither are higher education cross-curricular skills, the grounds for a discussion are beyond the scope of this chapter. In addition to these skills, it is considered that students also acquire specific skills derived from the following learning objectives:

- Understand the passive elements of the horizontal subsystem in a Structured Cabling System (SCS).
- Install UTP (Unshielded Twisted Pair) cabling and patch panel (patch panel) in the frame of a communications cabinet (rack).
- Perform installation and connection jacks RJ-45 network in user workspace.

- Make connections in the corresponding patch panel to the horizontal cabling.
- Naming and correctly label the installation carried out following general criteria.
- Check and verify correct operation of subsystem wiring.
- Understand the process for error handling in the network layer on the TCP/IP stack (Transmission Control Protocol/Internet Protocol).
- Experiment with real network error handling for IPv4 (IP Protocol, version 4).
- Schedule sending/receiving IP packets.
- Interfering with a TCP/IP network via the ICMP protocol (Internet Control Message Protocol).

This list of learning objectives is not exhaustive because the goal is for students to discover them for themselves. A more detailed elaboration of these can influence the student's work by implementing behaviorist learning methodology, rather than a constructivist approach [26].

4. Development of Activity

The application of PBL methodology in the context of a Degree in Telematics Engineering will be described through a case study of all the subjects involved. Each case study is presented independently but follow the same phases of development:

- Phase I. Problem Statements
- Phase II. Research.
- Phase III. Development of the project.
- Phase IV. Presentation of the solutions/results.

4.1. Team Workbook

Common material for the three cases has been used during PBL phases. Highlights include the use of team workbooks throughout the project (Table 1).

Table 2 shows an example of the operating rules provided as guidelines in the template team workbook. This information can be modified by each team to develop their own rules. Also teachers for each subject can modify the template to suit their needs. Similarly, the review of the team workbooks depends on each project. In some cases there is only one final delivery of the workbook and in other cases several partial deliveries are carried out.

Table 1. Section of template workbook team

Group nº:	Roles/Functions	Absences
Surname1 Surname2, Name	Representative (fixed)	
Surname1 Surname2, Name	organize the meetings (rotary)	
Surname1 Surname2, Name	divide task in groups (rotary)	
Surname1 Surname2, Name	Write documents (rotary)	
Observations:		
Operating rules of group		
List of rules and commitments acquired by the group members		
Consequences or compensation for non-compliance with the above rules		

Table 2. Examples of equipment's function rules

Team function rules (it can be written rules which every member accepts in order to be able to follow them)
Assist meetings planned by the group, practical classes, unless a justified reason exists Be responsible, as an individual and collectively, forthe work undertaken Be prepared to help others if they haven't finished their task Do not arrive late to meetings
If anybody doesn't follow any of these rules more than three times, they will be punished by completing an extra task If any of these rules aren't followed more than 6 times, that person will be expelled If anybody arrives late to the meeting, they will have to pay 50 cents, which will be used to benefit the group If anybody doesn't do their task, they will have to bring his laptop in following meetings

4.2. Rubrics and Assessment Questions

Rubrics for the assessment of group work in the different phases of PBL projects have been used. We have chosen a sufficiently verified tool [27-28] in which rubrics are used to establish uniform, consistent and transparent criteria to evaluate and grade the assignments used to evaluate students. The rubrics have been shared for the three case studies presented in the chapter, and have been designed to assess the following types of activities: group work, oral communication and written communication. The criteria established for each of the rubrics are evaluated according to a Likert [29-30] scale (4/evaluation louder -1/lower valuation) as shown in the example in figure 1, which shows an extract of the oral communication assessment rubric.

	1	2	3	4
1.—Exposure of the technical solution	The technical solutions is very poor.	The technical solutions is poor in general, but some technical solution can be spotted.	The technical solutions is good, but some question are unsolved.	The technical solutions are very good.
2.—Speech Organization	Technical solution are exposed directly, without anticipating even number.	Only the number of technical solution are expected.	They anticipated and classified the arguments or ideas, but does not summarize during exposure scheme.	They anticipated and classified the arguments or ideas, which are taken over and/or graphically support.
3.—Relevance and clarity of the argument	The arguments are not presented or not really related with the technical solution.	The arguments correspond to the technical solution but some of them are irrelevant or misunderstood.	The arguments are relevant but they are briefly exposed or with low explicitness.	The arguments are relevant and clearly expressed (specifying the assumptions, conclusions and justifications).

Figure 1. Example of rating scale used in the rubrics.

The assessment rubric for group work is structured around twelve criteria or indicators that are named below: preparation, contributions, job sharing, attitude, problem solving, focus on work, use of time, quality of work, interdependence, construction of group reflection and self-assessment and leadership. These indicators were chosen based on the application the section should serve both to assess the student's own work within the group (self) as well as the work of other members.

The main objective of the written assessment rubric for communication is to evaluate the final report submitted during the PBL activity. These reports contain a description of the solutions or work developed to overcome the problem or complete the initial project. This rubric consists of the following eight evaluation criteria: objectives, methodology, presentation of results, discussion of results, quality of presentation, writing, use of language and literature.

The assessment rubric for oral communication is used in all those activities in which the working groups present information to other groups, even though their progress in developing the project during intermediate stages is evident in results shown in the final phase. These presentations are made together, all group members participate, and it is supported by a software tool for presentations. This rubric is based on eight indicators of assessment: objectives, methodology, presentation of results, discussion of results, quality of presentation, writing, use of language and literature.

Finally, to evaluate the suitability of the PBL methodology in each subject, a quiz has been designed with fifteen questions grouped into the following seven dimensions (categories):

1. Contextualization.
2. Satisfaction and Motivation.
3. Learning.
4. Skills.
5. Improvement.
6. Teachers.
7. Methodology.

5. Case Studies

5.1. Case 1. Connection and Check UTP Cat 5e/6 on a Structured Cabling System (SCS)

5.1.A. Introduction

This project is carried out in the context of Networking Fundamentals subject. This subject belongs to the first semester of the Degree in Telematics Engineering. This project is developed within the activity called supervised work or ECTS tutoring.

This activity aims to install a structured cabling system pilot so that students understand the passive elements, learn about tools and techniques for connecting a patch panel and user network jacks as well as taking measurements and basic checks for verifying the correct operation of the cabling system.

5.1.B. Organization

The project is divided into four phases that will be presented later. It takes place over seven consecutive weeks, although the number of weeks varies according to the number of groups trained and the distribution of hours over the course. It consists of classroom group work sessions with the teacher for a total of five hours. There will be distance group work in phases II and IV of the project to develop preliminary documentation about the theoretical fundamentals and to prepare oral presentations.

5.1.C. Materials and Resources

The teacher provides students with a guide containing the statement of the project and implementation phases (see figure 2).

In that document the criteria and assessment tools that are used during the activity are also included. In addition, it offers workgroups a template team workbook as well as a calendar with important dates of the project. A repository of useful links about the theoretical fundamentals, regulations and standards associated with SCS is also provided. Finally, there is another set of links intended to advise on techniques of group work and oral presentations. All this information is available through the virtual space of the subject in the virtual campus of the university. In addition, each workgroup is provided with the materials and tools needed for the construction of a pilot SCS.

5.1.D. Phases of Development

Detailed description of the work done in each of phases of the project:

a) Phase I. Problem Statements

The first phase takes place in the classroom where the teacher has one hour to explain the PBL methodology to all workgroups. Previously workgroups have been established and communicated to students through a virtual campus forum. Then the teacher introduces students to the problem that the work to be performed is contextualized, including a brief explanation of the SCS and the EIA / TIA-568-B standard. Considering how the work will develop in the next phase, they are informed of the resources available to gather information. Finally, supporting resources on how to work together and make effective oral presentations are presented. Please note that these students have little or no previous experience in group work. Among the resources

Networking Fundamentals. ECTS tutoring: Problem Based Learning. 2014-15

CONNECTION AND CHECK UTP CAT 5E/6 ON A STRUCTURED CABLING SYSTEM (SCS)

Index

1. Defining the Problem

Considers that forms part as a network technician Department of Information and Communications of a business organization, whose main activity is the development of digital educational content aimed at primary school.

Management has decided to create a new space for demonstrations of new materials both commercial department, educational institutions, and other potential customers for their products.

This classroom is an open space of about 36 m2 in the semibasement floor (accessible from the ground floor), previously used as a warehouse for printed material. The same is provided with the technical means adequate lighting and air conditioning, but not available in any network infrastructure. However, it needs to provide network connectivity to the following equipment:

- 8 places for desktop computers.
- Up to 14 places for laptop computers.
- 1 place for desktop computer teacher.
- 1 place for an application server.

Management asks the department to provide a solution that classroom appropriate network infrastructure with the following requirements:

1. Physical network infrastructure must be integrated into the Structured Cabling System (SCS) of the company.
2. Must support network traffic on the LAN up to 1000 Mbps.
3. Must be flexible to allow connection to the network equipment such as PCs, laptops or IP phones.
4. Must have its own electrical panel with properly sized circuits to service computers in the classroom, including a circuit ready for connection of a UPS (Uninterruptible Power Supply) Local.
5. Design[1] a system for copper wiring unshielded twisted pair (UTP) Category (CAT) 5e/6 that meets the needs of connection under previous requirements.

[1] No purpose of this activity is the design of SCSs. Therefore the student is provided with a design of the distribution system and subsystem elements horizontally through the Fig.1.

Figure 2. Example of statement and working guide.

available to support group work, special emphasis is placed on the use of a team workbook, explaining in detail what it does and how it should be used throughout the entire project.

b) Phase II. Research

Development of this phase is by distance except the last part where workgroups should make an oral presentation of a wiring plan. It lasts four weeks during which workgroups should meet to organize the work to be performed and record your performance in the workbook team. In this phase, each working group will review the basic literature provided by the teacher. In addition, each workgroup may include other references as long as they are satisfied with respect to the reliability of the information source. From these documents, each workgroup should prepare a wiring plan which must include the following minimum information:

- Description of a Structured Cabling System (SCS).
- Description of EIA / TIA-568-B standard with respect to SCS.
- Description of materials and tools to be used.
- Description phase wiring connection.

To end this phase, each workgroup must present to other groups the wiring plan developed during a classroom session. These presentations will be evaluated both by the teacher (item 1 evaluation criterion C), and by the other groups (item 2 assessment criterion C), ending with a discussion/sharing for exchange of views between the different groups.

c) Phase III. Development of the Project

In this phase, each group will go to the network lab to install and connect an SCS pilot. This phase comprises between one or two weeks depending on the number of groups and the available material, with each group given one hour to complete it. The teacher will provide the material and tools already specified above for development of the following tasks:

1. Performing a user socket, consisting of connecting 2 CAT 5e jacks RJ-45 (with allowance T568-B) and its placement in the corresponding surface boxes.
2. Placing a CAT 5e patch panel in a mini rack 19" and connecting the horizontal cabling UTP CAT 5e (with allowance T568-B) that comes from the user sockets.
3. Verifying the installation using a wiring tester or microscanner.

Initially, these tasks must be performed without help from the teacher. Each group will use documentation gathered and developed for the wiring plan to carry it out. At the end of the session, the teacher can help groups to solve certain problems or overcome certain tasks. Once each pilot SCS is completed, the teacher will evaluate the final result, which corresponds to point 1 of criterion D of the final evaluation of this activity.

Finally, each group will prepare a project report that complies with the template provided. They must also analyze the final product and compare it with the pre-planning that was done in the previous phase, emphasizing deviations or successes in their conclusions. The teacher completes the assessment of this document, which is also part of the criterion D of the final evaluation.

d) Phase IV. Presentation of the results

In this phase workgroups must present to other groups the results and conclusions of the project under the following conditions:

- The minimum contents of which will be comprise:

    - Description of materials and tools used.
    - Development phase wiring.
    - Development of the verification phase.
    - Description of the SCE pilot developed and contextualized by the project.
    - Conclusions.

- Maximum time of 10 minutes with a maximum of 15 slides.
- A computer presentation tool should be used.

In the evaluation of presentations both teacher and the other groups participate (see the evaluation criteria E). This phase will end with a discussion/sharing to exchange views between the different groups.

5.1.E. Evaluation Criteria

The evaluation of this activity is based on five criteria, which include general criteria such as attendance and the student's contribution to the group throughout the activity, as well as the specific evaluation of the results for phases II, III and IV. To carry this out the rubrics already presented in section 4 are used. For a more detailed description of the assessment criteria:

A. Attendance and group activity sessions (up 1 point): This is a general criteria for assessing the activity. To assess this section attendance is recorded by the leader of each phase and should be included in the computer workbook. The following scale applies:
- 0 absences = 1 point.
- Between 1 and 2 absences = 0.5 points.
- 2 or more absences = 0 points.

B. Work done by the student within their group (up to 3 points): Supported by the same rubric, the student, as their peers should evaluate the attitude and own performance in relation to the work done within the group. The score co-evaluation of their peers is obtained as an average of all evaluations completed. In turn, the final score for this criteria is obtained as an average between points 1 and 2. The idea is to get as balanced an evaluation as possible between how the student has worked and how they have perceived by their peers. Therefore, the assessment criteria are:
   1. Self-assessment of the student.
   2. Co-evaluation of the student by the group.

C. Presentation of wiring plan (up 2 points): specific criteria to assess the outcome of the group in phase II. The method for calculating the score for this criteria is the same as that used for criteria B, for:
   1. Assessment by the teacher.
   2. Assessment by the other groups.

D. Production by the group (up to 2 points): This is specific criteria for the work completed during phase III. This is a technical and practical development which is the main objective of this activity, the assessment will be carried out exclusively by the teacher. For this criteria, the score is weighted for each assessment item as follows:
   1. Quality of the work done for the SCS pilot (75%).
   2. Evaluation of the final report (25%).

E. Presentation of results (up 2 points). Specific evaluation criteria for Phase IV where the corresponding rubric will be used. The score calculation will be completed the same way as the criteria for B and C:
   1. Assessment of the teacher.
   2. Assessment of the other groups.

5.2. Case 2. Error Handling in IPv4

5.2.A. Introduction

The development of the PBL activity in the Computer Networks course is made in lessons of 2 hours long for both theory and practice. In a normal week, two lessons are given and the project is performed over several weeks during the course.

5.2.B. Organization and Methodology

During the different life cycle of the project, each group self-manages its workbook, in which all activities are written down. In order to achieve manage it correctly, each group makes their own internal operation rules that, at least, must include the following items:

- Number of meetings.
- Designation of a secretary in the group to summarize the conclusions of each meeting.
- To decide the disciplinary measures that are going to be taken if a member of the group doesn't carry out a mandatory task.
- The development of the sessions in the classroom during phases II and III will follow this the following order:
- Each group presents their progress in a short presentation to the rest of the students.
- The teacher will participate only if necessary.
- Only one of the members of the group, to be chosen randomly by the teacher just before the explanation, will give the explanation.
- Each of the exhibitions will be evaluated both by the other groups and by the teacher.

5.2.C. Materials and Resources

In order to complete the activity, the teacher offers the students the necessary infrastructure to develop the project. For example, the Virtual Campus (a Moodle based platform) is used to create a forum thread for each group. This way, each group can discuss and share some information that is stored in the Moodle system. Other tools are also used such as DropBox. In this tool, the documentation necessary to develop the project can be shared in a public or private way.

5.2.D. Phases of Development

Next, a detailed description of the work completed during each phase of the Project is given.

a) Phase I. Problem Statements

The presentation of the problem in one session but in two parts. First it is necessary to explain the methodology that will be followed during the project. Then, the specific project that needs to be solved during the activity is also presented. Next, working groups are created. Each group is formed by three of four students. Therefore, the basic actions, lasting approximately 30 minutes each, are as follows:

- Explain the learning methodology to be used during the project.
- Explain the problem to be solved during the project.
- Create the working groups.
- Plan how to start the work.

b) Phase II. Research

In this phase, each group performs some investigations about the topic of the problem by reviewing and studying the literature. At this time, the teacher only offers some indications about the topic but the students need to search and learn by themselves.

At the end of this phase, each group will develop its own documentation, indicating how they are going to solve the problem, establishing and planning meetings and the work that is going to be done. Regarding this specific case study concerning error handling in IPv4, the key to the problem is in the IPv4 header, which can be shown in figure 3. As we know, IPv4 does not control errors, and we only found a checksum field in its header that only covers an error in its own IPv4 header.

After some years of experience with this problem, we have noticed that students conclude at this stage that IPv4 protocol is not aware of control errors and it needs other protocols to be able to do this. In fact, most of the groups choose to solve this problem by implementing the ICMP (Internet Control Message Protocol) protocol using the packet driver tool.

At the end of this phase, the students need to complete some tasks outlined in the Moodle platform:

- Documentation of the solution
- Workbook

<table>
<tr><th colspan="2">4-bit</th><th colspan="2">8-bit</th><th>16-bit</th><th colspan="2">32-bit</th></tr>
<tr><td colspan="2">Ver.</td><td colspan="2">Header Length</td><td>Type of Service</td><td colspan="2">Total Length</td></tr>
<tr><td colspan="5">Identification</td><td>Flags</td><td>Offset</td></tr>
<tr><td colspan="3">Time To Live</td><td colspan="2">Protocol</td><td colspan="2">Checksum</td></tr>
<tr><td colspan="7">Source Address</td></tr>
<tr><td colspan="7">Destination Address</td></tr>
<tr><td colspan="7">Options and Padding</td></tr>
</table>

Figure 3. IPv4 header.

c) Phase III. Development of the Project

This is the most time consuming phase during which each group performs their own tasks to achieve "handling errors in IPv4 packets". There are one hour meetings every two weeks so the teacher can follow the evolution of each group. The numbers of non-face meetings are managed by each group and are written down in the workbook.

At this time the PBL project, the Virtual Campus was used (tools such as forum, documentation, workshop and survey). We highlight the capacity of the tool workshop to enable the co-evaluation of the different phases. This tool also allows each group to upload the workbook and the final report for this phase, which includes the design of the work completed in this stage. To summarize, by the end of this phase each group has finished the main work for the project. During Phase III, each group presented the following:

- Documentation of the problem
    - Design of the solution
    - Implementation of the solution
- Workbook
    - Initial and final workbook
- Presentation of the solution
    - All groups need to deliver the presentation file that will be used in the final phase.

d) Phase IV. Presentation of the solution

As was shown in the previous phase, the complete documentation is presented during the output of Phase III. Thus, at this stage the work is orally presented. The mechanism for doing this is explained next. All members of each group must take part in the presentation, but the order of the speakers is randomly selected just before the presentation.

For each group, an evaluation by the rest of the students is carried out based on the Evaluation Criteria to be explained in the next section.

5.2.E. Evaluation Criteria

In order to evaluate this next activity, several factors are considered: Technical knowledge, participation in the activity, the quality of the final documentation, assessment by the rest of the students as well as self-assessment.

During each presentation, the teacher evaluates the evolution and the improvement of the solutions offered by each group. The oral presentation is also evaluated.

A rubric is used to evaluate the implementation of the final solution.

5.3. Case 3. Advanced Communications

5.3.A. Introduction

In this section another case study is presented. In this case, the project is about bandwidth networks and their capacity to offer Quality of Service (QoS) and Traffic Engineering (TE) in data communications. The goal of the project is to design an innovative networking solution for a company that has particular needs in terms of both throughput and delay guarantee, making it necessary to use quality of service and traffic engineering solutions. The main professional skill developed in this subject is leadership. This skill was selected to be taught in this subject because it is developed in the fourth year of the degree. In the last year of the degree, the students must understand the impact of their decisions, the consequences of leading a group and the importance of guiding your team in the right direction.

5.3.B. Organization

In this context, the learning activity has been developed as part of one specific topic of the subject, Quality of Service, bandwidth networks and its relation with IP networks.

The PBL methodology used to develop the aforementioned skills has been developed in different scenarios, such as: in classroom with the teacher, in a classroom without the teacher and through individual work. Using these scenarios as required combined with correct training, learning is reinforced through this PBL activity.

Regarding this specific case study, and taking into account that the entire group is usually formed by approximately 15 students, this kind of activities can be undertaken properly. In this way, it is possible for the teacher to listen for a long time and give instructions only if the groups are disorientated. This situation greatly improves the learning process because the teacher is aware of everything that is happening with the working groups.

5.3.C. Material

The material required to complete the PBL activity is a communications laboratory with enough capacity for the students (15 - 20 seats in our case). In this lab there are also some communications devices and other hardware that we are going to use to develop the project. One of the goals is to deploy the solution in this hardware so the students can test their design in real devices. This fact is very important for the project, because the students can compare their solutions with those proposed by the rest of the groups, and learn from them.

5.3.D. Development

As in the previous case studies, the development of the activity follows the same "Phases" structure. Next, we describe briefly these phases in the context of this case study.

- Phase I: Understanding the problem: This is an activity made in class. It is used to describe both the problem and the tasks required to develop the project. Moreover, if necessary, some concepts are introduced.
- Phase II: Investigation and autonomous work (3-4 hours long): Once the students have understood the problem, the groups are created (3/4 students per group) and they start developing the project. The following tasks are completed:
  - For one hour, the students look for the protocols and technologies that they need to study as well as the software and hardware needed a priori to solve the activity.
  - Next, the group brings their investigations together.

PBL Activity. Outline:

We have just joined to the Communications Department of a Company whose main activity is the organization, management and delivery of training courses both on-line and on-site.

The department consists of other 3 engineers. Our Company has been struggling with the performance of the network, which is a traditional TCP/IP network. The increase of customers in the last yeas has made to the raise number of failures in the network so, it collapses frequently, which means that there had many customer complaints.

The company board request us to design solutions in order to offer quality of service in the network and how we could improve performance and reliability. Furthermore, the volume of traffic is not monitored and they need we detail the maximum capacity that our network can support without a deterioration in the service provided. The solution should be written and detailed in a technical report with the results obtained from a testbed of 6 routers. An overview of the network is shown in the figure:

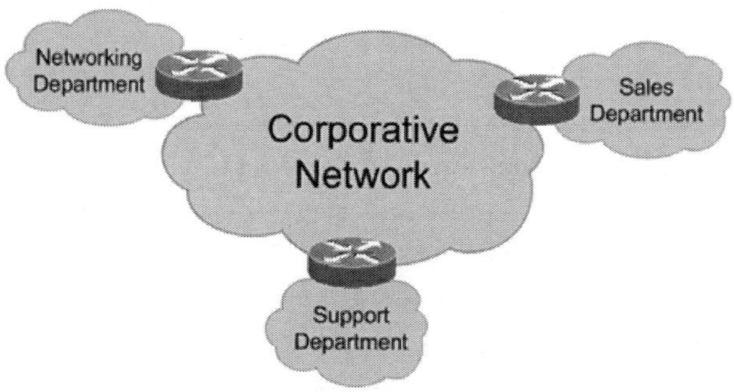

The requirements are:

- The Company has three departments: Sales, Support and Networking. The data traffic flows among the three departments following the behaviour explained next:
  - VoIP traffic: Among the 3 departments.
  - Video streaming: Between the networking department and the support department. This traffic is the most important because the operation of the company relies in the delivery of the on-line courses.
  - Backup: Both Sales Support department sends regularly the backups to the Networking department.
- The Budget to spend in hardware is limited to 8000 euros.

In this scenario, you need to:

- Identify the necessity of QoS in backbone networks and the limitation of IP networks to offer it.
- Find the hardware needed to deploy the project.
- Design and configure the testbed
- Plan and run the benchmarks.
- Analyze the results and prepare some conclusions

Figure 4. Outline of the PBL activity.

- The students also plan for the whole project. This planning will be revised regularly in order to detect any deviations from the scheduling and if so, corrective actions must be taken.
- With respect to the investigation phase, it is necessary to find relevant and reliable information to use in order to grasp the concepts and technologies needed to understand the problem.

    - Next, enough time is given to find the hardware needed to implement the solution.
    - Finally, the group summarizes the information obtained and the work that has to be done out of the classroom is decided.
- Phase III: Problem resolution. This phase is completed in the classroom and is very time consuming. With the decisions taken in the previous phase, each group develops the activity and tries to solve it. The documentation is used to write down the solution through an iterative process.
- Phase IV: Describe the solution and final evaluation. Finally, it is necessary to finish the project and to present the final report with the documentation where the solution is detailed. In addition, final conclusions are included.

In this case, one of the skills taught is the planning of an entire project. Initially, the students do not know how much time is necessary to finish because they have not worked with the technologies related to the project. In this situation, all groups are will deviate from the initial planning. These deviations will make the students implement the corrective actions that need to be taken to correct the deviations from the plan. In last phase, the initial planning is compared with the results obtained at the end of the project.

Figure 4 shows the outline of the project given to the students at the beginning of the activity.

6. Student Valuation of Academic Achievement

6.1. Case Study 1

The assessment of this activity has focused on two aspects: (1) compliance with the learning objectives for the development of the teamwork skill, (2) accomplish the general aim of the project for developing a technical work on a physical layer network model. Assessment of teamwork skills (1), this was done by an anonymous survey in which each student valued the teamwork skills in general. This survey was conducted at the end of phases II, III and IV to have an overall picture of the evolution of the groups throughout the life of the project and an overview of the activity.

Figure 5 shows the average rating obtained for each teamwork category assessed during phases II, III and IV. This is interesting because these are the phases in which students had to work together autonomously most of the time.

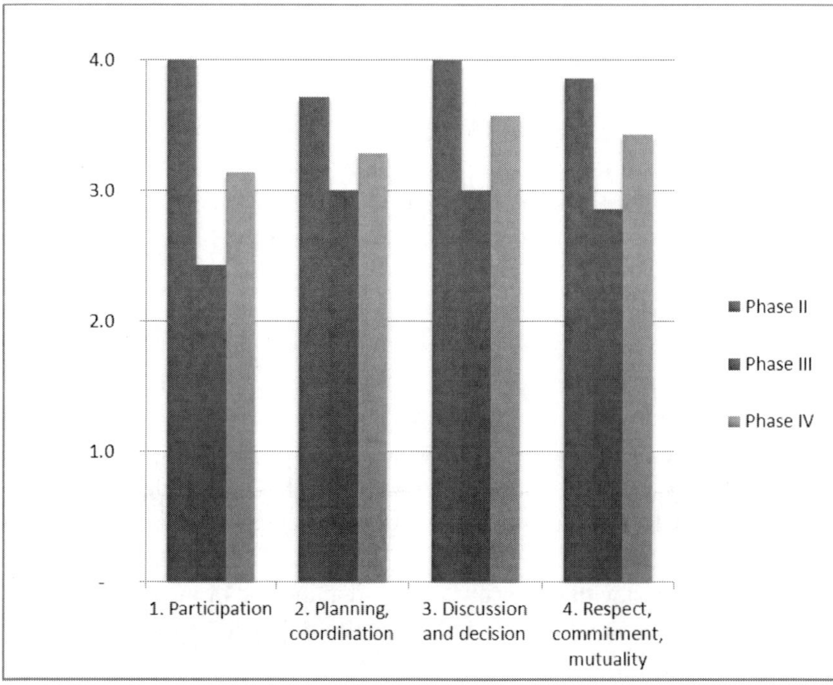

Figure 5. Average rate for teamwork skills during the project.

Based on the above, the following conclusions are drawn: during phase III, which corresponds approximately to half the lifetime of the groups, a lower score is obtained and a greater range of responses within the categories evaluated. This could be due to a natural process in which relationships within the group mature. So, a clear evolution in the views of the student throughout the life of project can be seen. The students had higher expectations for Phase II and real conception of teamwork during Phase IV. Moreover, most students value positively or very positively all categories during all phases of the project, the category "Discussion and decision" (with 81% satisfactory answers) and "Respect, commitment, mutuality "(with 76.2% satisfactory answers), followed by "Participation" (with 66.7% satisfactory answer) stand out. Finally note that only 14% believe it has been a waste of time after completion of the project.

6.2. Case Study 2

With respect to the second case study, we focus on the results in oral communication. As we described previously, in this project each group had to present its solution and to present it orally to the rest of the students. After that, the students were evaluated both by teachers and by the rest of the students. Figure 6 shows the numerical results obtained by the evaluations of the teachers.

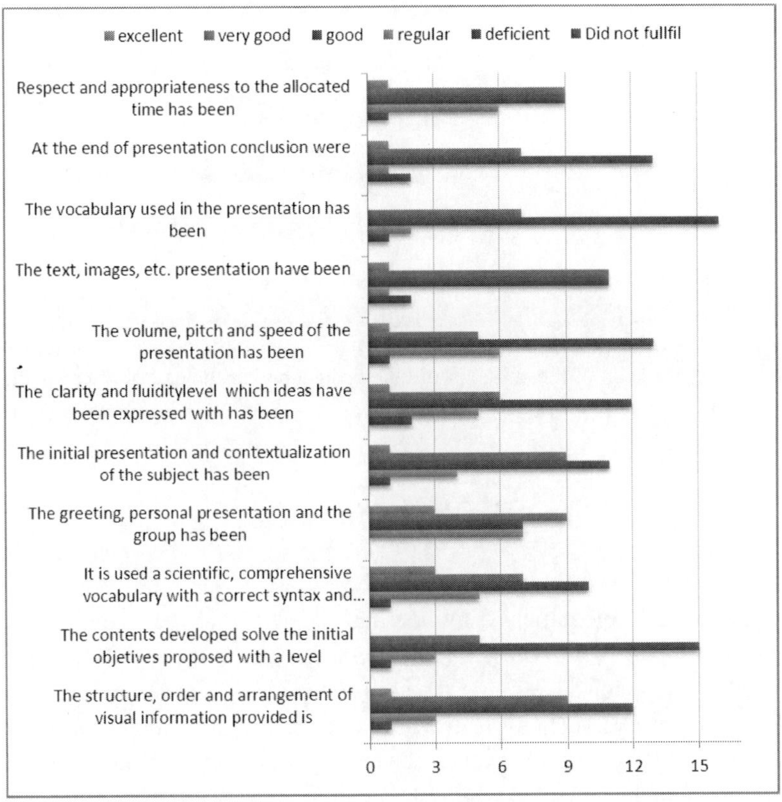

Figure 6. Numerical results of the oral communications skills in case study 2.

6.3. Case Study 3

In this case, we show the results obtained in the activity by a comparison of the skills valuation at the beginning and at the end of the project. Due to the nature of this project, where the students are mature (they are enrolled in the last course of the degree), we could teach some skills such as: time planning, self-confidence, teamwork, flexibility in the way of thinking and leadership. Figure 7 shows the results of the assessment from the students' point of view by a comparison of the skills before the project and at the end of the project. The valuation by the students was made through an anonymous questionnaire, with numerical questions from 1 = very poor to 5 = very good.

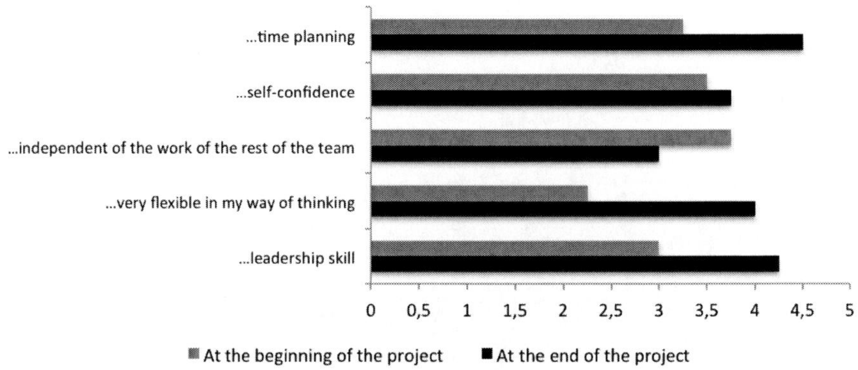

Figure 7. A comparison of the skills before the project and at the end of the project.

CONCLUSION

In this chapter we have described three case studies of Project Based Learning in a Telematics Engineering degree at the University of Extremadura, Spain. From these three experiences, it can be seen that PBL is a good methodology for developing the skills needed by students of engineering in the XXI century, such as teamwork, oral communication or leadership. Numerical results from each experience demonstrate the benefits for students of working in such activities, where they can learn in a different environment that is more exciting and more motivating than traditional learning. Moreover, during the activity and after it, both students and teachers have highlighted their satisfaction with this method, related closely to the type of work that students will undertake when they begin their career after finishing university. It is worth mentioning the extra effort that this methodology causes for teachers, as this means it is suitable for environments where the number of students is not very high.

REFERENCES

[1] Barrows HS, Tamblyn, R. *Problem-Based Learning: An Approach to Medical Education.* New York: Springer, 1980.

[2] Barrows HS. How to Design a Problem-based Curriculum for the Preclinical Years. New York: Springer, 1985.

[3] Barrows HS. The Tutorial Process. Springfield, IL: Southern Illinois University School of Medicine, 1988.

[4] Barrows, H. S. (1996), Problem-based learning in medicine and beyond: A brief overview. New Directions for Teaching and Learning, Volume 1996, Issue 68, pages 3–12, Winter 1996. DOI: 10.1002/tl.37219966804.

[5] P. del Canto, I. Gallego, J. M. López, E. Medina, F. Mochón, J. Mora and M. Valero. Folllow-up and feedback processes in the EHEA. *Journal of Technology and Science Education* 1(1), 2011, pp. 12-23.

[6] J. Rojas, X. Prats, A. Montlaur, M. Valero and E. García-Berro Rojas. Effectiveness of problem based learning for engineering curriculum. *Cases on Technological Adaptability and Transnational Learning*, 2010. 333-359. IGI Global.

[7] Juan F. Valenzuela-Valdés, Marina Aragón-Romero International "Patented! Problem based learning" *Conference on Education & E-Learning Innovations ICEELI'* 2012 , July 1-3, Sousse, Tunisia.

[8] Barrows, H. S., Schmidt, H. G., & De Volder, M. L. (1984). Tutorials in Problem-based Learning. A New Direction in Teaching the Health Professions. (null, Ed.) (Vol. null, p. 16).

[9] Adelsköld, G., Aleklett, K., Axelsson, R., & Blomgren, J. (1999). Problem-based distance learning of energy issues via computer network. *Distance Education*, 20(1), 129–143. doi:10.1080/0158791990200110.

[10] Greening, T., Kay, J., Kingston, J. H., & Crawford, K. (1996). Results of a PBL Trial in First-year Computer Science. In Proceedings of the 2Nd Australasian Conference on Computer Science Education (pp. 201–206). New York, NY, USA: ACM. DOI:10.1145/299359.299388.

[11] Reeves, T. C., & Laffey, J. M. (1999). Design, Assessment, and Evaluation of a Problem-based Learning Environment in Undergraduate Engineering. *Higher Education Research & Development*, 18(2), 219–232. DOI:10.1080/0729436990180205.

[12] Waters, R., & McCracken, M. (1997). Assessment and evaluation in problem-based learning. In Frontiers in Education Conference, 1997. 27th Annual Conference. Teaching and Learning in an Era of

Change. Proceedings. (Vol. 2, pp. 689–693 vol.2). DOI: 10.1109/FIE. 1997.635894.

[13] Northwood, M. (2003). Problem-Based Learning (PBL): from the health sciences to engineering to value-added in the workplace. Global J. of Engng. Educ., 7(2). Retrieved from http://198.66.238.102/ journals/GJEE/Publish/vol7no2/Northwood.pdf.

[14] Hassan, A., Ariffin, M., & Yusof, K. M. (2004). A review and survey of Problem-Based Learning application in Engineering Education. In *Conference on Engineering Education* (pp. 14–15). Kuala Lumpur. Retrieved from http://eprints.utm.my/975.

[15] Hsieh, C., & Knight, L. (2008). Problem-Based Learning for Engineering Students: An Evidence-Based Comparative Study. *The Journal of Academic Librarianship*, 34(1), 25–30. doi:10.1016/j.acalib.2007.11.007.

[16] Mills, J., & Treagust, D. (2003). Engineering Education – Is Problem-Based Or Project-Based Learning The Answer? *Australasian Journal of Engineering Education.* Retrieved from http://www.academia.edu/ download/31039433/Engineering_ Education.pdf.

[17] Perrenet, J. C., Bouhuijs, P. A. J., & Smits, J. G. M. M. (2000). The Suitability of Problem-based Learning for Engineering Education: Theory and practice. *Teaching in Higher Education*, 5(3), 345–358. doi:10.1080/713699144.

[18] Brodeur, D., Young, P., & Blair, K. (2002). Problem-based learning in aerospace engineering education. In Proceedings of the 2002 American Society for Engineering Education Annual Conference & Exposition. Retrieved from http://www.dept.aoe.vt.edu/~cdhall/courses/AUAE/ Brodeuretal.pdf.

[19] Dahlgren, M. A., & Dahlgren, L. O. (2002). Portraits of PBL: students' experiences of the characteristics of problem-based learning in physiotherapy, computer engineering and psychology. *Instructional Science,* 30(2), 111–127. DOI:10.1023/A:1014819418051.

[20] Kay, J., Barg, M., Fekete, A., Greening, T., Hollands, O., Kingston, J. H., & Crawford, K. (2000). Problem-Based Learning for Foundation Computer Science Courses. *Computer Science Education*, 10(2), 109–128. DOI:10.1076/0899-3408(200008)10:2;1-C;FT109

[21] Linge, N., & Parsons, D. (2006). Problem-based learning as an effective tool for teaching computer network design. Education, *IEEE Transactions* on, 49(1), 5–10. DOI:10.1109/TE.2005.852600

[22] Polanco, R., Calderón, P., & Delgado, F. (2004). Effects of a problem-based learning program on engineering students' academic achievements in a Mexican university. *Innovations in Education and Teaching International,* 41(2), 145–155. DOI:10.1080/14703 29042000 208675

[23] Said, S. M., Adikan, F. R. M., Mekhilef, S., & Rahim, N. A. (2005). Implementation of the problem-based learning approach in the Department of Electrical Engineering, University of Malaya. *European Journal of Engineering Education*, 30(1), 129–136. DOI:10.1080 /03043790512331313895.

[24] Striegel, A., & Rover, D. T. (2002). Problem-based learning in an introductory computer engineering course. In Frontiers in Education, 2002. FIE 2002. 32nd Annual (Vol. 2, pp. F1G–7–F1G–12 vol.2). DOI:10.1109/FIE.2002.1158138

[25] Chickering, A.W. y Gamson, Z.F. (1987). "Seven principles for good practice in undergraduate education", *American Association for Higher Education Bulletin*, Marzo 1987

[26] Ackoff, R. L., & Greenberg, D. (2008). Turning learning right side up: Putting education back on track. Pearson Prentice Hall.

[27] Andrade, H. G. (2000). Using rubrics to promote thinking and learning. *Educational Leadership*, 57(5), 13-19.

[28] Arter, J., & McTighe, J. (2001). Scoring rubrics in the classroom: Using performance criteria for assessing and improving student performance. Corwin Press.

[29] Likert, R. (1932). A technique for the measurement of attitudes. *Archives of psychology*.

[30] Likert, R. (1974). A method of constructing an attitude scale. Scaling: a sourcebook for behavioral scientists. Chicago: Aldine, 233-243.

In: Project Based Learning on Engineering ISBN: 978-1-63482-217-6
Editors: J. Valenzuela-Valdés et al.

Chapter 4

PBL ON INFORMATION AND COMMUNICATIONS TECHNOLOGY (ICT)

Juan C. González-Macias, Pedro J. Pardo and Juan F. Valenzuela-Valdés***

Department of Computing and Telematics System Engineering,
Universidad de Extremadura,
Centro Universitario de Mérida, Mérida, Spain

ABSTRACT

This chapter discusses the application of the PBL methodology on Information and Communications Technology (ICT). First, a small collection of various works focusing on ICT through the PBL methodology is detailed. Then, the overall project which implements the PBL methodology at the *Centro Universitario de Mérida* is presented. Subsequently, three case studies of the PBL methodology implemented in three subjects at the University of Extremadura are described and analysed. The first case is an innovation project conducted on the subject of mobile wireless networks, in which the students have developed a professional project to install and configure a WiFi network. In the second case study, students use their mobile phones to make a set of measurements of the mobile phone networks in an area of the city. The

* E-mail juanvalenzuela@unex.es.

third case study presented is a description of how this learning methodology is implemented on the subject of Standardization and Telecommunication Projects, where the students undertake the design of an ICT project (Common Telecommunications Infrastructure). Thus, the students are encouraged to search optimal solutions, which reduce the consumption of materials to be installed. Another part of the project consists of studying home automation solutions which reduce energy consumption of buildings, with the aim of designing more sustainable buildings.

1. Introduction

The current information and communication technologies (ICT), whose development is unstoppable and expansive in all the fields in recent years, are also starting to be applied on university teaching methodologies. This is the case of the methodology of Problem Based Learning (PBL), a work technique for small groups in which a group of up to eight students develop, produce and resolve in a fairly autonomous way, a problem with the contents of the subject or degree. This is a technique that emerged in the 60s of last century in North American universities and health sciences studies. It requires group sessions of face to face interaction, so apparently, it is far from virtual technologies designed for the contact of students in different places, either synchronously (in real time, such as chat or video conferencing) or asynchronously (e.g. email or forums).

Nevertheless, these new technologies have also been introduced in the PBL methodology since the late 90s of the last century. This chapter briefly reviews the main applications and solutions for online PBL, where some of the authors in this issue describe specific teaching experiences.

2. Global Project Based on Information and Communication Technologies (ICT) at the C. U. M.

Skills development through active learning methodologies is suitable for environments in which it is required specific professional training by students. A group of teaching innovation formed by the authors of this manuscript is developing a teaching innovation framework, whose aim is to develop certain practical skills related to ICT (which are useful for their graduates) through active learning methodologies.

Teachers have the common goal of innovating when conducting their classes. For this reason, they have decided to develop this teaching innovation project based on their previous experience [1-2]. The project is designed in two phases: the first phase took place during the academic course 2012/2013 and the second phase during 2013/2014. In the first phase, each teacher implemented in their subject an innovation related to Information Technology. Specifically, the subjects involved are:

- Subject 1: Data Transmission. Problem to solve: Making antennas in the 2.4 GHz band to improve the coverage of WiFi networks.
- Subject 2: Wireless Mobile Networks. Problem to solve: Performing coverage maps for wireless networks.
- Subject 3: Regulations and Telecommunications Projects. Problem to solve: Creating an ICT project (Common Telecommunications Infrastructure) for the administrative building of the *Centro Universitario de Mérida*.

In the first phase lecturers should develop a project by making innovations in the area of Information Technology. In the second phase of the global project developed in the academic course 2013-2014, all projects were linked as a puzzle, generating a large project as a result of the combination of the work in several subjects. Therefore, a map of the projects of the degree is obtained, where the final objectives of acquisition of skills related to information technology were achieved, through a sequence of projects.

Consequently, the main objective of the project is to create and consolidate a group of teachers who work continuously and systematically improving the teaching and the application of Information Technology and Communications.

2.1. Analysis of the Current Situation

This project should be fitted within the subjects taught in the degree of Telematics Engineering. This degree provides access to a regulated profession, Telecommunications Engineer, which enables the free exercise of the profession based on their professional responsibilities. The motivations to start this educational project can be describe as follows:

1) Universities have traditionally been well-considered to teach knowledge, but not for training for a professional career, the know-how, as highlighted in the project [3]. This certainly requires a reflection on the teaching-learning methods currently employed in most universities [4-5]. The process of the adaptation to the European Higher Education Area (EHEA) is understood to be a pretext to start a change in the teaching-learning model used in universities [6]. As it is indicated in the framework for the Spanish University System Integration in the EHEA [7], the new university model must be based on the following points:

    a) Introduction of the European credit (ECTS), as the academic unit which measures the overall student work.
    b) The curricula and teaching schedules must be designed from the academic and professional profiles demanded by society.
    c) The educational objectives have a professional orientation, integrating the basic generic skills, transversal skills related to the integral development of individuals and specific skills.
    d) The organization of the degrees is focused around learning, using active methodologies where the student is the fundamental item.

2) On the other hand, Europe demands new telecommunications professionals, highly skilled, internationally oriented and responsive to the growing needs of citizens, businesses and administrations. These engineers, with a great capacity to provide advanced and complex technical solutions, should be effective, sustainable, accessible and close to society. Therefore, the training received by students is demanded to be more practical every day. In addition, the consultant recruitment "Adecco Professional", in his report of the most sought professional in 2011, are looking for professionals with initiative and ability to work in teams. The latter two skills can be obtained through active learning.
3) Currently, the average number of students in the Degree of Telematics Engineering at the *Centro Universitario* of *Mérida* is about 20. This allows the implementation of active learning methodologies that require small workgroups.
4) Last but not least, among the objectives of the Degree in Telematics Engineering is important to cite: "Skills to solve problems with initiative, decision making, creativity and ability to communicate and

transmit knowledge, understanding ethical and professional responsibility of the activity of Telecommunications Engineer", i.e. students which obtain the degree should have acquired these skills.

Analysing the current situation, the changes that are happening in teaching, the demands of companies and the objectives and skills listed in the Bachelor degree in Telematics Engineering, it is needed to provide students with the ability to solve problems of the real life and motivate them to do it efficiently. Therefore, it was decided to implement an experience whose aim is the acquisition of the skills, abilities and competencies by the students, which are required in their professional career (including the capacity for continuous learning). Additionally, the student is responsible for his own learning. The active learning methodology and project-based learning (PBL) is applied for carrying out this ambitious task. Therefore, this group of teachers, in line with the Bologna process, considers that the student is the focus of the educational system and actively participates in the design, evaluation and development at all levels.

2.2. Goals of the Project

The main objective of our teaching project is to strengthen the acquisition of skills related to Information Technology and Communications through an active learning methodology as PBL. The student will have to carry out various tasks for each subproject, in order to obtain the following skills:

- Developing critical thinking
- Being able to make decisions
- Being able to work in teams.
- Troubleshooting.
- Planning your work.
- Having a broad knowledge of Information and Communication Technologies.

The specific objectives derived from achieving this main objective are:

1. Having a catalogue of activities described in the PBL methodology which develops innovative issues related to Information Technology and Communications. This information could be used by any team of teachers.
2. Providing tools to work with generic skills in addition to technical skills.
3. Moving from a teaching methodology where the kernel is the teacher, to a teaching methodology where the kernel is the student and its learning process.

The planning is developed jointly by the entire teaching group, following the methodology described in Section 2.4 for each project. All individual projects will have an individualized planning for each subject. Moreover, the use of the "subject forum" reinforces the knowledge of the student thanks to the online advice, within the course available for each subject on the platform of the virtual campus. Thus, a "community of learning, sharing and good practice" will be created giving the possibility to "live" the experience of the learning process through the continued use of Information Technology. Therefore, the current proposal of teaching improvement makes use of the ICTs, since, in our opinion, it is considered fundamental in current education systems.

3. Project 1 in the Subject of Wireless Networks

3.1. Introduction

The subject of wireless and mobile networks is an optative subject taught in the fourth year of the degree of Telematics Engineering at the *Centro Universitario* of *Mérida*. There is a great number of wireless networks laboratories in the world. Firstly, there are laboratory courses that use simulation software. These e-learning labs provide some advantages over conventional engineering laboratories. The simulation laboratory consists of a simplified version of a system. Although criticized for not incorporating enough real-life experiences, they have some advantages for the learning process. In this group, some university learning plans have been developed specifically for teaching with simulation tools. As a sub-group, it is possible to cite WiFiSiM from the University of Huelva (Spain) [8] and WiFi Lab from the University of Hong Kong (Hong Kong) [9] and a very recent revision of this laboratory type can be found in [10]. Another type of simulation tool used by Universities is "OMNeT ++", an extendible, modular, component-based C ++ simulation library for building network simulators that is used at the Universities of Budapest (Hungary) [11] and Monash University, (Australia) [12], among others. The last subgroup includes a great number of Universities that use communication tools developed on Matlab-Simulink in their laboratory courses, such as the John Hopkins University [13]. Whatever the sub-group, it is essential to take into account the software licenses available and whether these tools can be installed on the students' computer. Additionally, it is important to know if it is mandatory to use the software in the laboratory.

In a second group, there are laboratory courses that include some measuring tasks or the use of hands-on RF systems. For example, at Stanford University [14], students learn how to use spectrum and network analyzers combined with some simulation tools. In addition, some laboratory courses offer the possibility of designing and building different parts of a wireless system or network. For example, at Texas University (U.S.A) [15], the laboratory is based on Lab View and RF hardware.

3.2. Pedagogical Aspects

The learning objectives to be obtained with the completion of these projects are shown in Table I and Table II, where the skills and the level of development of these skills (1 minimum - 5 maximum) to be acquired for the students are displayed.

Table I. Learning Objectives

Learning Objectives
1. Knowledge of the different standards for wireless networks.
2. Knowledge of the different simulation tools for WiFi networks
3. Knowledge of the theoretical aspects of WiFi networks
4. Knowledge of the necessary equipment to create a WiFi network
5. Knowledge of several software measurements tools for network WiFi
6. Knowledge of the different measurement procedures WiFi networks
7. Knowledge of the different tools to measure WiFi networks

Two different sections in the field of wireless communications have been described. The project-based learning (PBL) methodology has been used in both parts. To complete each of the sections, students had to perform a report where the project is described. Additionally, students had to defend their work and each of the decisions made throughout the project in the classroom in front of their classmates.

Table II. Skills to be acquired

Skills	*Sub-Project 1*	*Sub-Project 2*
Self learning	5	4
Creative thinking	3	5
Taking the initiative	4	4
Development of critical thinking	4	4
Ability to work in group	4	4
Communication Skills	4	4
Troubleshooting	3	3
Data analysis	4	4
Oral presentations	4	5
Reporting	4	4

Next, the following sections are described:

3.3. Design of a WiFi Network

In this project, students combine the PBL methodology with web platforms. The project lies on designing a WiFi network for the *University of Extremadura* and presenting a professional project suitable of being verified by the relevant engineering association. Additionally, students will obtain the professional skills required for a telecommunications engineer.

In Figure 1a, a screenshot of one of the platforms of the project is shown.

3.3.A. Coverage Map of a WiFi Network

In this project, students work with the PBL methodology in combination with the use of a professional software. The project involves making a coverage map of the various existing WiFi networks in our Centre. Students have been asked to identify all the WiFi access points in our Centre and place them on a plane. Additionally, they should identify the areas of maximum speed and their association with the several technologies presented in the centre. The areas with interference problems should also be located and identified on a plane.

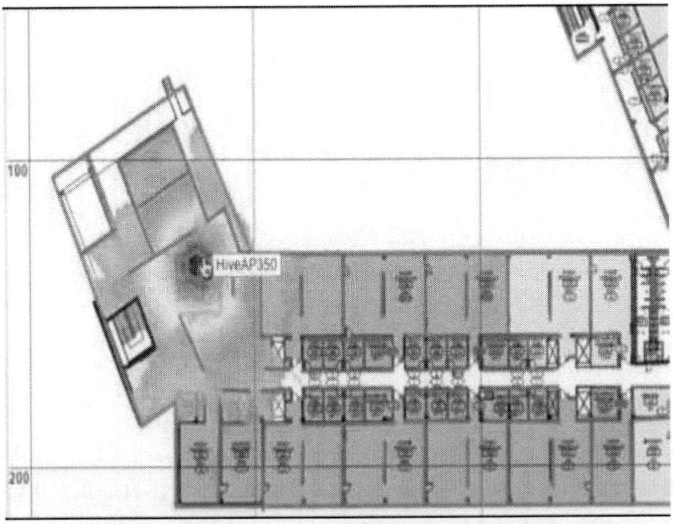

(a)

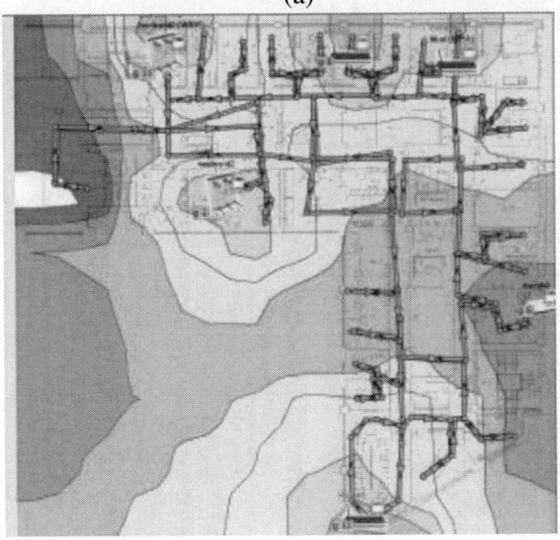

(b)

Figure 1. Screenshots. (a). Web platform which carries out the simulation and computation of the network coverages. (b). Professional software for measuring coverages of WiFi networks.

To begin the project, students have to install the professional software called Ekahau on their laptops (http://www.ekahau.com). In Figure 1b, a screenshot for this software is displayed. Students are also provided with maps of the different rooms of the center to locate the access points.

3.4. Results

Two kinds of results can be distinguished in this project:

1. The results of the project, i.e. the degree of acquisition of the learning objectives and skills proposed, explained in paragraph 2.
2. Evaluation of the project, i.e. the degree of satisfaction with the project by students and teachers.

3.4.A. Project Results

Two strategies have been applied to determine the degree of acquisition of the learning objectives: the evaluation of the submitted and defended reports and the test of the knowledge gained. The results were very satisfactory, since over 90% of the students were shown to have acquired more than 50% of the learning objectives.

3.4.B. Project Assessment

The project was evaluated by the students through a survey in which they were asked about various aspects of the project. The average rating was 4.1 out of 5, which reflects a high degree of satisfaction. For the teachers, the performance of the project has been successful since their students have been highly motivated and have achieved great results.

4. Project 2 in the Subject of Regulations and Telecommunication Projects

4.1. Introduction

The subject of regulations and telecommunication projects is an optative subject taught in the fourth year of the degree of Telematics Engineering at the *Centro Universitario* of *Mérida*. After passing this subject students should be able to work as a self-employed applying the acquired skills. Among the different job opportunities for the degree, one is the free exercise of the profession. Therefore, the most common project among of different possibilities of Telecommunication projects to be performed by graduates in Telematics Engineering is the Common Telecommunications Infrastructure (ICT).

In order to carry out the first ICT project, it has been considered to start with the most common site for all students, which is the administrative building of the *Centro Universitario of Mérida*, where they study.

4.2. Learning Objectives

Through this activity students are expected to achieve the following learning objectives:

- Ability to learn independently.
- Facing a real problem and being consistent with the decisions to be taken.
- Starting a project previously nonstudied technologies.
- Participating in a team with similar experts.
- Preparing brief technical meetings.
- Justifying technical solutions or proposals.
- Discussing, listening to other opinions and accepting the consensus by the team.
- Leadership groups (understand the concept of leadership).

In short, students are expected to learn to work in a group of experts looking for a technical solution to a real problem and collaborating together.

4.3. Planning

The first day of class, the teacher explains what PBL is and how learning takes place. Students are grouped into groups of at least 3 students, and maximum 4 students.

The ICT project consists of different parts, namely: services of television, telephone, broadband through coaxial cable, twisted pair cable and optical fiber; home automation, and infrastructure that supports all these telecommunications services. The teacher distributes to each group the development of one of the parts of the ICT project and insists on finding optimal solutions that minimize the amount of material to be installed.

Each group delivers some digitalized plans of a part of the work at the end of the course, so that the ICT project is carried out by the whole class.

Furthermore, the groups should establish several internal rules of each group. These regulations should include, at least:

- The minimum number of internal meetings held by the group.
- A member of the team should act as a secretary, in order to collect a summary of the issues raised and solved in each group meeting.
- Disciplinary measures. When a member of the group does not attend the meetings, arrive late, or does not perform the assigned tasks, this member could be expelled of the group.

The design of an ICT project is an ambitious task which students wonder if they will able to develop with no knowledge at the beginning of the course. It's an ambitious idea and a big challenge for them.

The structure, numbering and characteristics of the plans to be delivered are described in the Royal Decree 346/2011.

4.3.A. Guidelines to Perform the Activity: Checkpoints

The teacher should provide the required materials for the subject, which are available to the students in the virtual platform of the University. Among the information, it is remarkable:

- Where to find educational information for the performance of the work.
- Planning of the subject, including the delivery dates of each of the parts of the project.

During the first two weeks of the course, the teacher explains the most important parts of the ICT project to be performed by the students in groups. The development of the remaining classes is completed as follows:

- In each hour of class, the progress of one of the groups is outlined.
- The exposition of each group is performed by using the classroom projector.
- A member of the team, chosen randomly, explains the decisions made in the current version of the project.

These meetings, in which the teacher participates as a listener, are checkpoints of the tracking of the activity.

4.4. Evaluation Criteria

The report to be delivered is a diary in which activities and decisions taken by the group are recorded for each session. Originality and optimized solutions are assessed whereas the copy of other sources is penalized.

For the exposition of each group, the secretary delivers the part of the completed report, the process of each meeting held by the group, and its possible conflicts.

The following points are considered for the evaluation of the activity: acquired knowledge, participation in the activity, proactivity, technical quality, presentation of the final report, peer assessment and self-assessment. The classmates evaluate the work done by the group that carries out the presentation. All these parameters are evaluated according to a signature that outlines the subject criteria, which is known previously by the students. This signature places special emphasis on the parameters of the learning objectives.

In each of the presentations of the groups, the teacher writes down the evolution and improvement of the design solutions adopted for each group.

4.5. Results

Results can be divided into two parts: on the one hand, the achievement of the learning objectives and, on the other hand, the satisfaction of the participants in the project. According to the knowledge obtained by the students, the 92% of them developed and followed the projects, improving their teamwork skills and learning to create ICT projects, which can be developed in their real life as engineers. The scores, on average, were higher than 6.5 out of 10. 15% of the students failed. Finally, the student satisfaction was measured through surveys that students completed over various aspects of the course.

Conclusion

Firstly, a state of the art of the PBL implementation in university subjects related to ICT has been presented. Next, the authors have described the teaching innovation project related to PBL developed in the *Centro Universitario* of *Mérida*. Results show that most of the objectives have been satisfied. Students have developed all the projects with enthusiasm and motivation, mainly because these works are rather related to the real life. This educational experience has been very positive for the degree of satisfaction and involvement of all parties. Other universities may establish similar experiences making use of the concepts and situations which occur in the projects presented in this manuscript.

References

[1] Gil, J., Perez, A. L., Suero, M. I., Solano P. and Pardo, P. J. (2010) Evaluation of the Effectiveness of a Method of Active Learning Based on Reigeluth and Stein's Elaboration Theory, *Int. J. Eng. Educ.* 26, 628-641.

[2] Martínez, G., Pérez A. L., Suero M. I., Pardo P. J. (2012) The Effectiveness of Concept Maps in Teaching Physics Concepts Applied to Engineering Education: Experimental Comparison of the Amount of Learning Achieved With and Without Concept Maps, *J. Sci. Educ. and Technol.,* 5, 1-11.

[3] Universidad de Deusto & Universidad de Groningen (2003), *Tuning Educational Structures in Europe.* Bilbao: Universidad de Deusto.

[4] Santos Guerra, M. A. (1999), 20 paradojas de la evaluación del alumnado en la Universidad española, *Rev. Electrónica Interuniversitaria Formación del Profesorado,* nº 2

[5] Font Ribas, A. (2004), Las líneas maestras del aprendizaje por problemas, *Rev. Electrónica Interuniversitaria Formación del Profesorado,* nº 18

[6] Michavila, F. y Calvo, B. (2000), *La universidad española hacia Europa.* Madrid: Fundación Alfonso García Escudero.

[7] Ministerio de Educación, Cultura y Deportes (2003), *La integración del sistema universitario español en el Espacio Europeo de Enseñanza Superior,* Documento-marco. Madrid: Ministerio de Educación, Cultura y Deportes.

[8] Mateo Sanguino, T. J., Serrano Lopez, C., Marquez Hernandez, F. A., WiFiSiM: An Educational Tool for the Study and Design of Wireless Networks, Education, *IEEE Transactions* on, vol.56, no.2, pp.149-155, May 2013.

[9] Lin Cui, Fung Po Tso, Di Yao, Wijia Jia, The WiFi Lab: A Web-Based WiFi Laboratory Platform for Wireless Networking Education, Learning Technologies, *IEEE Transactions* on, vol.5, no.4, pp. 291-303, Fourth Quarter 2012.

[10] Gampe, A., Melkonyan, A., Pontual, M., Akopian, D., An Assessment of Remote Laboratory Experiments in Radio Communication, Education, *IEEE Transactions* on, vol. 57, no. 1, pp.12,19, Feb. 2014.

[11] Varga A. and Hornig R., An overview of the OMNeT ++ simulation environment, in *Proc. 1st Simutools,* 2008, vol. 60, pp. 1-10.

[12] Department of Electrical and Computer Systems Engineering, *Monash University Performance of Telecommunication Networks* available online: http://titania.ctie.monash.edu.au/netperf/ (Accessed: 16 Feb., 2014).

[13] *The Math Works,* Johns Hopkins University establishes virtual laboratory for wireless communications with Simulink, (Online). Available: http://www.mathworks.es/company/user_stories/Johns-Hopkins-University-Establishes-Virtual-Laboratory-for-Wireless-Communications-with-Simulink.html (Accessed: 16 Feb., 2014).

[14] Stanford University, *EE144/EE245 - Wireless electromagnetic design laboratory,* (Online). Available: https://ccnet.stanford.edu/ee144/ (Accessed: 16 Feb., 2014).

[15] The University of Texas, *EE 471C/EE 381V: Wireless Communications Lab* available http://www.profheath.org/teaching/ee-371c-ee-381v-wireless-communications-lab/ (Accessed: 16 Feb., 2014).

In: Project Based Learning on Engineering
Editors: J. Valenzuela-Valdés et al.
ISBN: 978-1-63482-217-6

Chapter 5

PBL AND GROUP TUTORING

A. J. Lozano-Guerrero[1,*], A. M. Martínez-González and A. Díaz-Morcillo

[1]Universidad Politécnica de Cartagena, Departamento de Tecnologías de la Información y las Comunicaciones, Plaza del Hospital, Cartagena, Murcia, Spain

ABSTRACT

In this chapter, the work carried out in an educational network program in a Spanish University is shown. This work is focused on using problem-based learning PBL jointly with group tutoring. Lecturers from several technical degrees have participated. Initially, it has been studied the collection of information by surveying students about the status of their University tutoring programs and how they think these programs should be carried out. Then, several activities to complement and reinforce the learning of the subject through PBL and group tutorials have been developed. Finally, we have evaluated the effectiveness of the activities by means of the final qualification of the subject and the degree of satisfaction of the students during the program through surveys.

---

* Corresponding author: A. J. Lozano-Guerrero. Universidad Politécnica de Cartagena, Departamento de Tecnologías de la Información y las Comunicaciones, Plaza del Hospital, 1. Cartagena, 30202 Murcia, SPAIN. E-mail: antonio.lozano@upct.es.

INTRODUCTION

Tutorials are an important tool for lecturers within the framework of the European Higher Education Area (EHEA) [1]. With the help of tutoring, lectures in the classroom and student self-learning are reinforced. Additionally, lecturers can obtain feedback information to follow the student learning process. Traditionally, there are three main models of tutoring: the French or the Napoleonic model, in which, learning is limited to acquiring technical knowledge, the Humboldtian model in Germany devoted to the scientific methods and techniques and finally the English one more focused on the training for a specific profession [2]. In the Spanish university, tutoring is usually identified with the activity that allows the acquisition of knowledge and skills. Although the student is provided with the tutoring tool, traditionally, in Spain and during the last years, tutoring visits are related to doubts to pass an examination. In many cases, these questions are directly related to the contents of the exam and not to a significant learning of the subject.

PBL was first introduced in McMaster University [3] in Canada in Medical degrees and has spread all over the world to other disciplines such as engineering. Currently, PBL is an abbreviation for both Problem-Based learning and Project-Based Learning [4]. We have developed activities for both problem solving or for longer project developments during the whole semester. It allows different degrees of teacher control or self-directed student learning. It can help students to develop critical thinking and problem solving skills. A real project or problem has to be solved. The students assume the responsibility of the learning and the lecturers act as facilitators. Hence, tutoring techniques become a powerful tool to develop this kind of projects.

It becomes necessary to know the starting point of the Spanish university, finding out what kind of tutorials have been developed traditionally and identifying the needs that arise within the framework of the EHEA to supplement the based on the lifelong learning (LLL) or learning throughout life.

In this chapter we evaluate the state of the tutorials at a public Spanish Technical University. Moreover, we have designed a general and comprehensive framework that individual teachers can adapt to their needs and use in PBL environments.

Basically, we can distinguish between two types of tutoring: the individual and the group tutoring. We are going to focus on the development of different activities of Problem/Project Based Learning in various technical courses to see how group tutorials can help in technical degrees and how they can be performed. The aim is to identify which are the desirable qualities in a teacher when carrying out these tutorials. It is necessary to know how to lead a group and how to apply collaborative techniques to help us in our work. Among these techniques, we find those cited in [5]. We have implemented these techniques in the activities developed during this study.

In [1] we find five of the personal qualities that a tutor must have considering the college students opinion: affection or capacity for empathy, personal knowledge of students, justice or fairness in dealing, quiet authority and respect for all students. Similarly, a lecturer should avoid attitudes such as paternalistic, inquisitive, crony, cold or functional and moralizing. Culture, age, abilities, personal circumstances, motivation, learning style, etc. are some facts to understand the current concerns of young people who also have different profiles. Therefore, in some cases it may be necessary to adapt the teaching method to the students' abilities, prior knowledge and learning objectives. In terms of desirable academic characteristics, the lecturer must be competent in the field under study and have skills of teaching methodology and collaborative learning.

In the Spanish law that is currently applied, 'Ley Orgánica de Universidades' (LOU) in Article 46, paragraph e) the right of students to receive "advice and assistance from teachers and tutors in the way it is determined" is recognized.

The factors that lead to the consideration of the tutoring activities are various: college is an institution changing continuously, the possibilities in the current university curriculum are many, there are a high number of academic failure cases and it is necessary to achieve maturity and student autonomy. In addition, the profile of the type of Spanish college students is changing. The lecturer must be able to act as facilitator of resources and learning tools. Nowadays, the lecturer is not just a transmitter of knowledge. All this happens within the context of the Information and Knowledge Society that requires constant training by teachers. Therefore, the EHEA University lecturer is expected to be able to carry out the role of tutor and advisor.

Previous studies have been carried out in [6-10]. This work has been developed in the technical degrees of telecommunication engineering and computer engineering. The authors of this paper are lecturers in the area of Signal Theory. The groups which we have worked with include all courses and grades so we have obtained a heterogeneous sample. The age of students ranges from 18 to 24 years.

There have been about two hundred surveys both in the initial and in the final stage from the last four academic courses.

In the survey also Management students have been included.

STAGES

The planning carried out by the lecturers includes the following sections:

Data Collection Stage

A survey has been carried out to obtain information on the status of the student's evolution in the degree, his experience with tutorials during their time at University, and the need for training in treating the student during tutoring by lecturers, the teacher's responsibility as tutor and the character of tutorials.

This information has served as a starting point to begin the work in the classroom.

Group Tutorials Stage

Each lecturer has created a set of tutorials addressing the needs of their course and defining the activities according to the most convenient criteria. To carry out these activities a common framework has been created for the lecturers involved in the experience. However, a high level of freedom has also been given to conduct the activity. The courses on which the activities were performed ranged from 10 to 60 students per course. The number of students who composed each tutoring group, therefore, has been generally limited between 2 and 30 students. The members of each tutoring group have been selected by the teacher randomly. Other options to create the groups were discarded for sake of brevity. The length of each group tutoring has been limited in all cases from 30 minutes to 2 hours per session. Concerning the location where the group tutorials have been carried out, although some teachers have chosen to conduct tutoring in the same office, others have decided to proceed with it in specific classrooms due to the size of the groups. From 3 to 5 meetings per semester for the tutoring group activities have been carried out. At least, a tutoring group session at the beginning and another one at the end of the experience.

The rest of the tutorials have been carried out for the development of the PBL activity. Group tutorials have been implemented to help in Project/ Problem-Based Learning activities. Projects were supervised throughout the entire semester and students having questions or doubts could easily solve them during these tutorials. The obligatoriness of the meetings attendance was decided by the lecturers. In some cases it was decided to use part of the time devoted to general tutorials. In connection with the evaluation and the importance that the group tutorials have had in the final qualification, in general, these have not been taken into account when grading the students in the final mark. The work carried out by the students has been evaluated as traditionally and a final exam was used to qualify the students as it was done until now.

Group Tutorials Assessment Stage

Finally, surveys have been conducted to evaluate the effect that the group tutorials had in the learning process of the students. These surveys assessed the work done by the lecturers concerning their guidance, personal attention, accessibility and use of resources and services. It has also been asked about the relationships developed by the students with their equals and the way these tutorials have helped in their social and academic skills.

RESULTS

When presenting the results no difference has been made in between academic courses. In Figure 1 we can see how 66% of students say that they normally attend tutoring while the rest usually do not. With regard to how often they attend, in Figure 2 it is shown that 22% never go to tutoring while 57%, the largest group, attend sometimes during the semester. 19% say they do it once a month and 2% once a week.

In Figure 3, we can see that the students recognize the importance of tutoring as only 5% believe it is not necessary to attend them while most of the rest of the students think it is necessary or at least desirable.

In Figure 4, the habits of attendance to tutorials are shown. Nearly 60% prefer to attend unaccompanied and 41% attend accompanied by a person. Only 2% of students usually attend tutoring within a group.

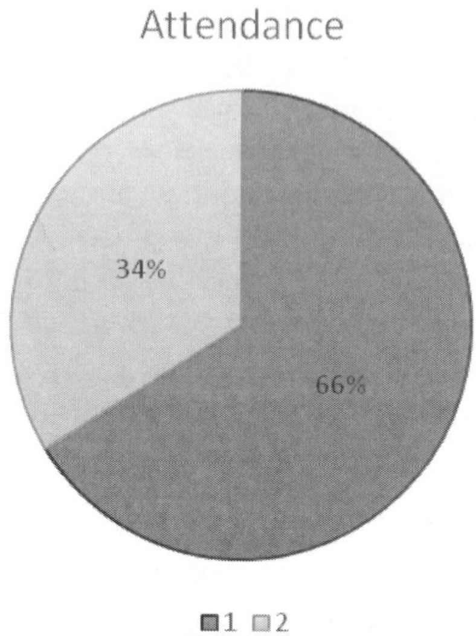

Figure 1. Attendance to tutorials 1- Yes. 2- No.

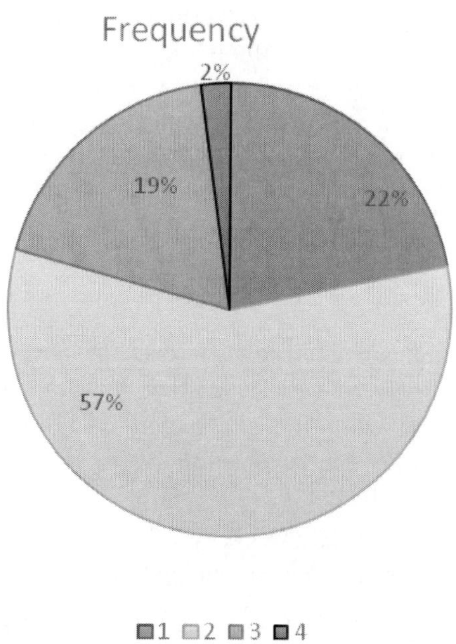

Figure 2. Frequency of attendance to tutorials. 1- Never. 2-Once during the semester. 3- At least once a month during the semester. 4- At least once a week.

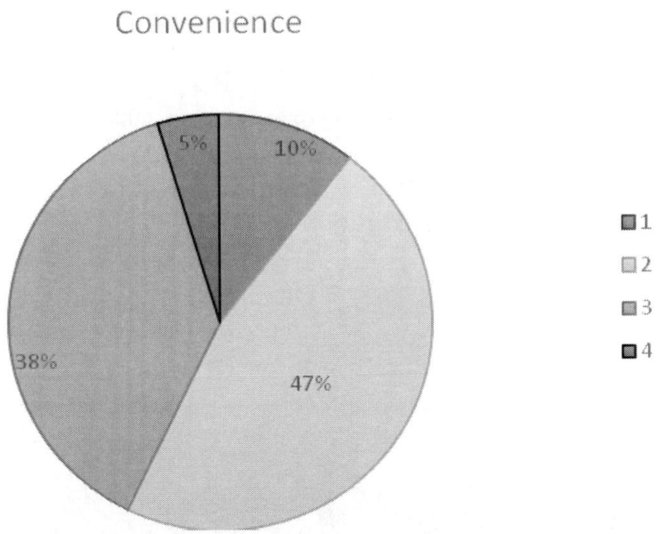

Figure 3. Convenience of attendance to tutorials. 1- Very necessary. 2- Necessary. 3- Convenient. 4- Null.

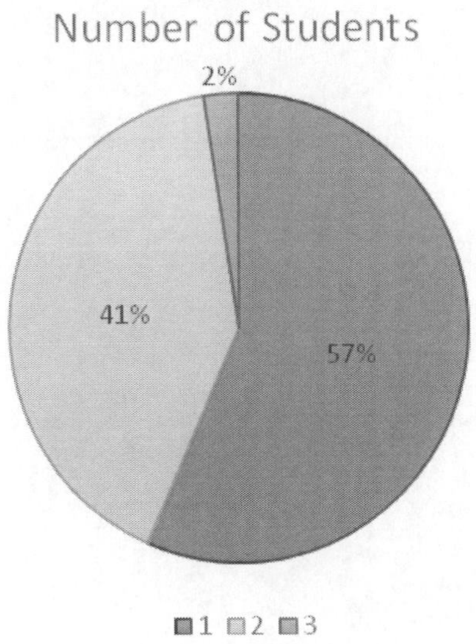

Figure 4. Assistance. 1-Unaccompanied. 2- With a partner. 3- In a group.

Figure 5 shows the impression of the students surveyed about the current state of the capacities of lecturers to work as a tutor and in Figure 6 the state that they think it would be desirable.

As it is seen, students believe that lecturers should be better trained in these tasks; consequently, their training is deficient in this aspect.

Figure 7 shows that the majority of the students tend to prefer tutorials that are purely academic or with professional interests which do not include personal aspects.

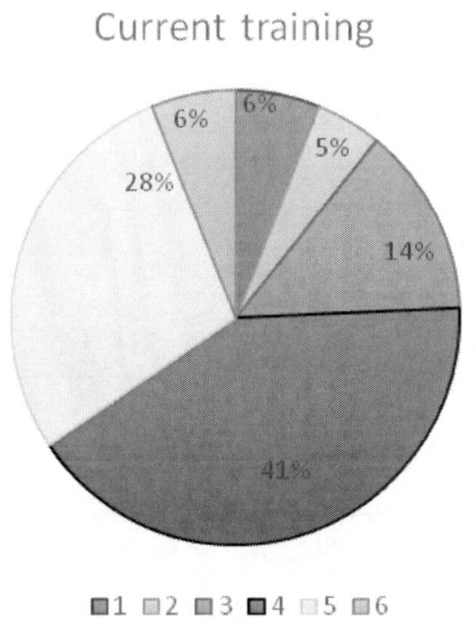

Figure 5. Current specific training on the task of tutor. Very poor 1- to 6- very good.

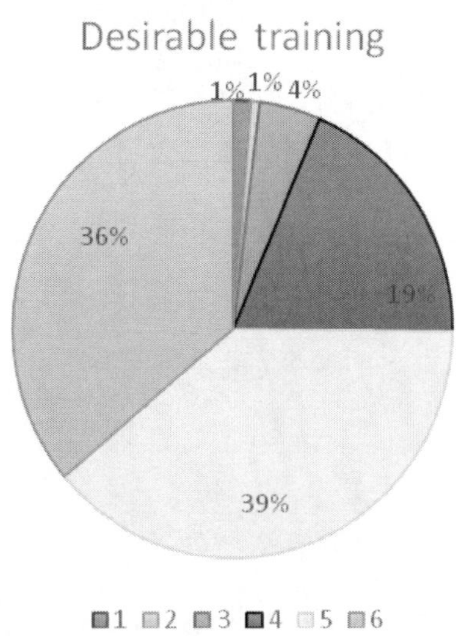

Figure 6. Desirable specific training on the task of tutor. Very poor 1- to 6- very good.

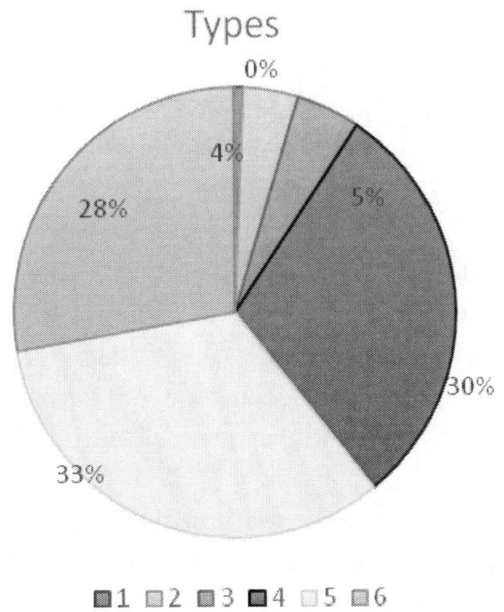

Figure 7. Types of tutoring. 1- Strictly personal -6- academic and professional.

Among the qualities considered to be the most desirable in a tutor, students choose, from highest to lowest, as follows: empathy, trust, understanding, acceptance and harmony with the youth world, the latter being the lowest rated largely below the rest.

The tutoring activities on PBL were well received by the students. After finishing them the subjective impression of the lecturers was positive. The grades compared to previous years, generally, were slightly higher but other factors may have influenced this fact. The final surveys were used to verify the adequacy of the activities undertaken. Although not mandatory, most of the students claimed to have attended all or nearly all sessions of group tutorials and only 7% affirmed to have never attended. Almost 100% of students claimed to have been informed of meeting times and were able to contact the lecturer when it was needed. The most widely used mean of communication was e-mail. Students stated that the planning of their studies improved. Similarly, the activities helped to improve the relationship between students. The students have not only learnt from each other but also from sharing knowledge and organizing themselves in the process of collaborative learning [4]. The satisfaction of the students when asking them was highly related to the inclusion of real life PBL activities. This raised their interest when comparing with the traditional Spanish way of lecturing, more based in an instructivist learning.

CONCLUSION

To conclude, students highly appreciated the inclusion of a PBL aided by group tutoring meetings. Students themselves believed that tutoring is convenient and necessary for learning and for real life and academic skills development.

Tutorials, according to the study, have not yet achieved the importance they deserve. It is necessary to improve the training of lecturers in tutoring tasks. The most valued tutoring qualities by students are empathy, trust and understanding. The less valued are acceptance and harmony with the world of youth. The preferred model is the academic tutoring versus the personal advice.

The tutoring activities have been well received and contributed to improve academic outcomes for students as well as planning studies and social skills.

REFERENCES

[1] A. Pantoja Vallejo and T. J. Campoy Aranda, *Planes de acción tutorial en la universidad*, Universidad de Jaén, Jaén, 2009.

[2] S. Rodriguez Espinar, Manual de tutoría universitaria. *Recursos para la acción*, Octaedro, Barcelona, 2004.

[3] D. Woods, *How to gain most from PBL.* Mc Master University Press, Hamilton, Ontario, 1994.

[4] X. Du, E. De Graaff and A. Kolmos, *Research on PBL practice in engineering education,* Sense Publishers, Rotterdam, 2009.

[5] E. F. Barkley, K. P. Cross and C. H. Major, *Collaborative learning techniques: A handbook for college faculty*. Jossey-Bass, 2005.

[6] A. J. Lozano-Guerrero, F. J. Clemente-Fernández, J. D. Sánchez-Heredia, J. F. Valenzuela-Valdés, J. Monzó-Cabrera, A. Martínez-González, J. Fayos-Fernández, and M. Dolores Cano-Baños, Las tutorías grupales en carreras y grados técnicos, *II Congreso Internacional sobre Aprendizaje, Innovación y Competitividad (CINAIC 2013)*, Madrid, Spain, November 2013.

[7] J. Monzó-Cabrera, A. J. Lozano-Guerrero and J. Fayos-Fernández, Tutoría grupal para la resolución de problemas de síntesis de amplificadores sintonizados y análisis de mezcladores en Ingeniería de

Telecomunicación, *I Congreso Internacional de Innovación Docente*, Cartagena (Murcia), Spain, July 2011.

[8] A. J. Lozano-Guerrero, J. Monzó-Cabrera and J. Fayos-Fernández, Las Tutorías grupales en los nuevos planes de estudio adaptados al Espacio Europeo de Educación Superior: Una experiencia en carreras técnicas. *I Congreso Internacional de Innovación Docente*, Cartagena (Murcia), Spain, July 2011.

[9] J. Fayos-Fernández, J. Monzó-Cabrera and A. J. Lozano-Guerrero. Ensayo de tutoría grupal mediante el método de resolución de problemas en asignaturas técnicas de primer curso adaptadas al EEES. *I Congreso Internacional de Innovación Docente*, Cartagena (Murcia), Spain, July 2011.

[10] A. J. Lozano-Guerrero and J. F. Valenzuela-Valdés, A Radionavigation Laboratory Course, *IEEE Transactions on Education*, Accepted for publication in Early Access, 2014.

In: Project Based Learning on Engineering
Editors: J. Valenzuela-Valdés et al.
ISBN: 978-1-63482-217-6

Chapter 6

ENHANCING E-LEARNING: PROBLEM BASED LEARNING SUPPORTED IN MOODLE

J. Enrique Agudo*, Mercedes Rico, Héctor Sánchez and Juan Manuel Vaca
Centro Universitario de Mérida, Universidad de Extremadura,
Mérida, Spain

ABSTRACT

Departing from the changing educational context we are moving in, Problem-Based Learning (PBL), combined with other teaching approaches, such as collaborative learning and e-learning, is an educational paradigm that promotes the development of valuable competences and enhances students' autonomy, reflection and team skills. In this paper we describe how the Moodle Learning Management System is being used to support the development of problem based scenarios in a higher education engineering subject, database analysis. The approach is implemented by combining tools of the LMS and specific plug-ins to implement six classical stages of PBL.

The implementation of the learning management system into a Problem-Based Learning environment offers the students engaging tasks and quasi real world contexts and provides them with situations they will work in as graduates and future professionals.

* Emails: jeagudo@unex.es, mricogar@unex.es, sasah@unex.es, juvasa@unex.es

INTRODUCTION

Online learning has considerably grown in the recent years both at user and technological level. So, there has been a shift in the way teaching is nowadays presented, opening new paths to online proposals, a real alternative for those demanding learning solutions without time or distance constraints.

Virtual learning environments or learning management systems (LMS) provide users with the technology for the development of online teaching, offering the objects and tools which are necessary for its effective implementation: access control, document repository, communication, assessment, etc. Though there are plenty of software for this and many other purposes, a good known alternative is Moodle, a free software e-learning platform created by Martin Dougiamas as an answer to proprietary platforms which were first used in colleges and educational institutions. Moodle is modular so that we can customize or extend it through modules aiming at adapting it to our educational needs. This characteristic along with its great widespread acceptance and open source nature are the main advantages of this platform.

Besides online demands, educational paradigms have also changed the traditional class dynamics into more collaborative learning environments, teaching proposals among which Problem Based Learning (PBL) stands out [1] with proven results in terms of developing learning, competences and skills [2].

Thus, learning Management Systems, following the path of modern Internet applications, are opening the door to new possibilities in collaborative learning, context in which modern LMS such as Moodle offers a wide range of possibilities that enable the integration of PBL and other collaborative paradigms in education [3]. Class or group forums, chat-rooms, wikis, databases or glossaries are a few examples of the tools that the teacher may use to support collaborative pedagogical approaches.

To cover our aim, the description of how the Moodle platform can be used to support the development of problem based learning in Higher Education, this paper is structured as follows: in section 2 we describe the pedagogical basis of our approach, subdivided into 3 subsections: (1) the pedagogical implications of PBLs (2) the description of Moodle platform and its main tools for this purpose and (3) proposals for PBL evaluation. Section 3 describes a practical online exploitation of PBLs at higher education whereas section 4 draws the final remarks and conclusions.

DEVELOPING PROJECT BASED LEARNING IN MOODLE

PBL and Transformative Learning

Educational paradigms are changing since learning is no longer seen as a way to acquire knowledge but to develop competences [4] such as the ability to work in teams, to manage time, solve problems, gain communicative abilities and critical thinking and the like, approaches which are changing the traditional class into collaborative learning environments in both face to face and in online learning.

Among this changing scenario, Problem Based Learning (PBL), defined in a wide sense as the process of working collaboratively towards the understanding or resolution of problems, is a teaching method comprising the completion of case studies and design projects which demand from the learner, among others, the acquisition of critical and problem solving skills. PBL provides a learning environment where students need to make decisions in quasi-real world scenarios and are encouraged to take the control of their own learning and apply knowledge at higher levels of completion [5].

In most cases, PBL is performed in small groups, fostering discussion and collaborative discovery in order to work-out the solution for a specific problem or learning situation. Educational literature has shown the benefits of using PBL and other approaches that promote active collaborative learning to improve students' thinking skills [6], processes which are pursued by tasks including:

1. A focus on real-world situations
2. An enquiry process
3. Acquisition of competences and skills
4. Projects supported and facilitated by teachers
5. Group-work and collaboration
6. Feedback and revision

According to [7], PBL is a cyclic process which comprises 3 main phases [8]: (i) cooperative thinking to identify initial problems and learning needs, (ii) individual self-guided study (iii) application of new acquired knowledge to given problems cooperatively and in a synthetic way. The PBL model has been applied in different educational environments, for different degrees and areas [9]. In this sense, [10] made a revision of traditional approaches and whose conclusions stated that PBL has positive effects in most cases and can be complemented by other teaching proposals. However, to arrange the agendas of all the members of a team to collaborate in work sessions is frequently a difficult issue, that is why the possibility of meeting remotely can allow more effective PBL initiatives.

In this sense, the use of the Internet to access learning materials, to interact with the content, with instructors and other learners and to obtain support during the learning process make possible the objective of knowledge construction. Our proposal on the application of PBL in e-learning situations is linked to the usage of the Learning Management Systems (LMS) Moodle as an application which promotes social communication and collaborative learning and offers a wide range of possibilities that enable the integration of PBL and other collaborative paradigms in class as we will describe below. Although Moodle capabilities to develop online learning and as a tool for managing learning are widely known, in this section we focus firstly on exploring the possibilities to develop PBL with standard Moodle tools, then we go through specific modules for this purpose and finally we present how to combine external tools to enhance Moodle's own resources.

Tools for PBL Development in Moodle

The focus in this section falls on standard Moodle tools; tools which will help us to plan the development of our project are easily identified. Moodle core does not incorporate specific PBL modules so the choices are adapted to the use of Moodle tools that simply serve to collect evidences/deliverables or specific PBL purpose modules integrated in Moodle mainly oriented to development of blended or traditional learning [11, 12].

Groups

The creation and management of groups in Moodle allows each resource or activity of the platform to be viewed or completed by groups of students. Besides, there exists the possibility that certain resources or activities are hidden to other groups. Groupings are groups formed by one or more established groups. This option allows us to limit the access to some types of tasks/ resources.

The simplest and most useful tools when working with PBL are the groups since each project will be developed by different groups of students. With the groups, students will join work teams in which they will exchange information and complete tasks through Moodle resources and activities. This action can be supervised by teachers throughout all process.

Forums

Forums could be the most important and most used tool in Moodle platform to manage projects. This tool can be used for brainstorming, debating, discussing and arranging meetings. In addition to that, forums are also used as a site of information exchange where links, images and videos can be added and any type of file uploaded. Forums are asynchronous since the participants do not have access to the system simultaneously. Users can be subscribed to the forum at any time and messages will be directly sent to the participants' mail when messages are posted.

If we use separate forum groups (we have to select the option "separated groups" from the forum settings) the students of each group will only see and reply to topics/threads of students belonging to the same group. Forums are the most used communication tool for e-learning. Likewise, this tool is an excellent communication tool to develop PBL so that each member of a group can talk to other members to develop their project.

Chat

Chats allow participants to have real-time synchronous discussion via text messages. This is an effective and useful way to make an online tutoring with a teacher or debating quickly between project members. Chats can store previous conversations even if one wants to use those conversations as an assessment method. Besides, chats can be used in separate groups so that each group has their own conversation.

Resources: Files, Pages and Direct Links

Files are resources teachers upload to the platform and students read and download. These files can be project scripts, images and sounds to use in projects, documentation related to the project, etc. Furthermore, teachers can create web pages or upload direct links to Moodle in order to provide information and documentation to students about the project. The function of these resources is similar to the role of files, but in this case external information can be included to Moodle if necessary.

These resources may also be separated by groups, for instance, if we want each group see their own resources then we will use groupings.

Assignments

Moodle assignments are exercises or deliverables proposed by the teacher in which students upload the text resolution or files. These tasks can have delivery deadlines. The best of Moodle assignments is that it allows using different types of assessment (seen in more detail in the following section). Assessments can be done by grading tasks with numerical rating or another rating scale, assessment guides or rubrics. Besides, all task assessments allow teachers to give feedback through an editor for each group of students.

In the case of PBL, designing a set of appropriate tasks (eg. one for each phase of a project) serves as a guide for students. So students can see the steps which they should follow when developing their project.

Conditional Access

Conditional access refers to resources or activities in which the teacher can indicate certain condition for the activities' availability to groups of students. Conditional access is aimed at guiding students through several steps and these students can only access to the step (resource/activity) if the deadline is finished, if they have completed a previous step or they have received a grade higher than the grade established in another resource/activity. These activities are very useful to perform projects steps by steps with several deliverables where teachers do not have to worry about making resources/activities available. Likewise these will be automatically activated if students go completing the requirements set by the teacher in the course.

Workshop

A workshop is a work team activity which allows students to participate in peer assessment. With this activity, teachers not only grade the assignment of each student but also the assessment they do of their peers. Interaction and collaborative work take place in a workshop mainly when cross-peer assessment is performed. In this phase, each student can learn how other students solve the same problem, improving their point of views and their learning. Besides, students should be critical with the work of others and grade them.

External Modules for PBL in Moodle

There are modules specifically designed for the development of PBL or other projects within the Moodle platform, these modules, created for this purpose, include a clear methodology and a variety of tools for the development of projects.

Techproject

This Activity Module is a collaborative constructivist workplace for developing project-driven learning activities using the standard IEEE specified "successive refinement" [13]. It provides tree-shaped friendly editors allowing writing and constructing the descriptions produced, reporting all steps of a project activity.

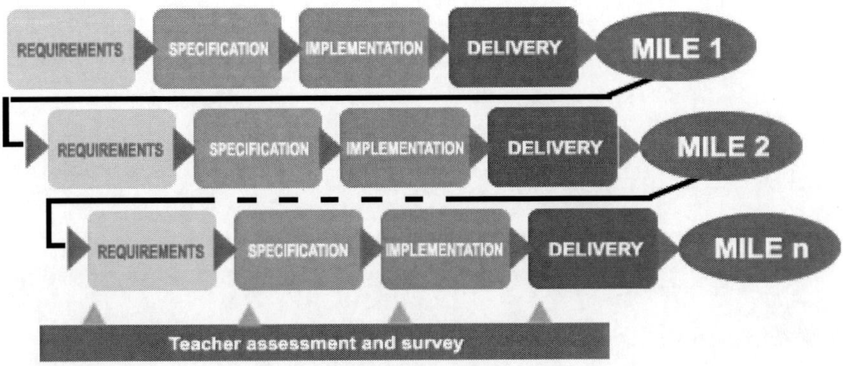

Figure 1. Project phases and iterative milestones [13].

Its description model is oriented towards "software development" project type with the corresponding phases (requirements, specification, implementation and deliverable) but could be adapted to whatever engineering project. The Techproject module is also suitable for iterative project cycles, where a project is performed defining milestones (Figure 1).

The module presents detailed information for the development of each of the project phases as well as for the management. It also provides tools for self-assessment and to test the work of individual students (Figure 2).

PBLTool

It is a block which implements a project management tool created to be used for collaborative learning. It is integrated with group forum and Moodle chat and its main function is to allow teacher and students to plan their activities together (Figure 3).

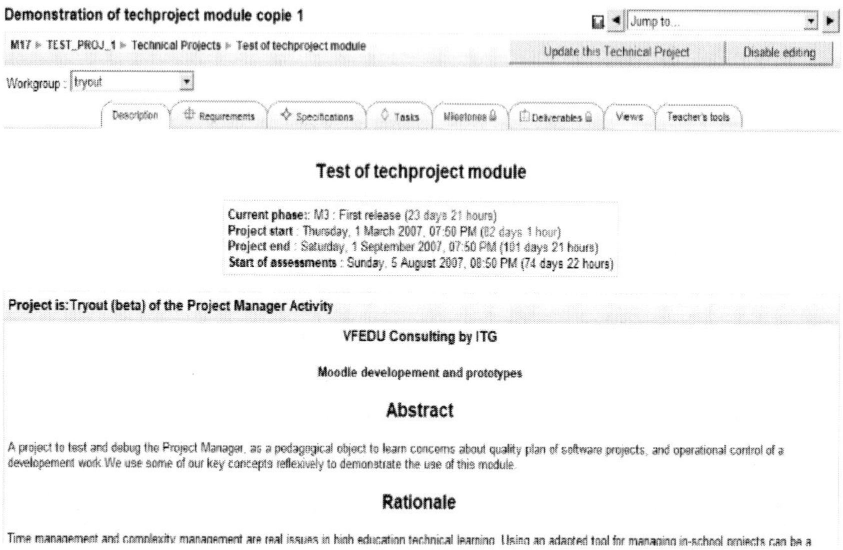

Figure 2. Techproject description screenshot [13].

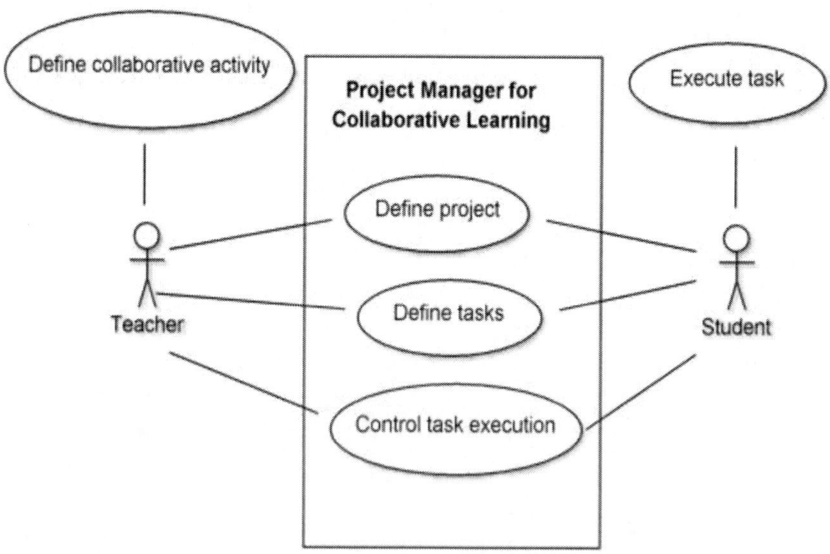

Figure 3. PBLTool diagram [14].

It allows managing students' assignments progression, managing groups and showing Gantt diagrams of the project to be carried out (Figure 4).

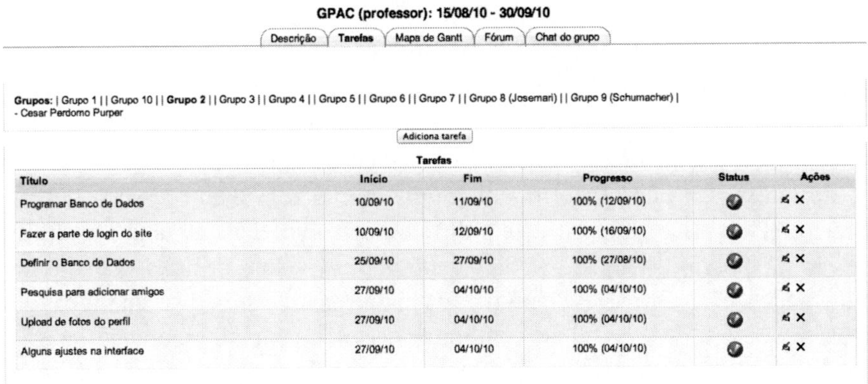

Figure 4. PBLTool description screenshot [14].

External Tools

There is also the possibility to use or link any external tool that will allow us to develop our project tracking. Among all the existing tools we present Mahara due to its integration with Moodle and Mural.ly by their collaborative nature.

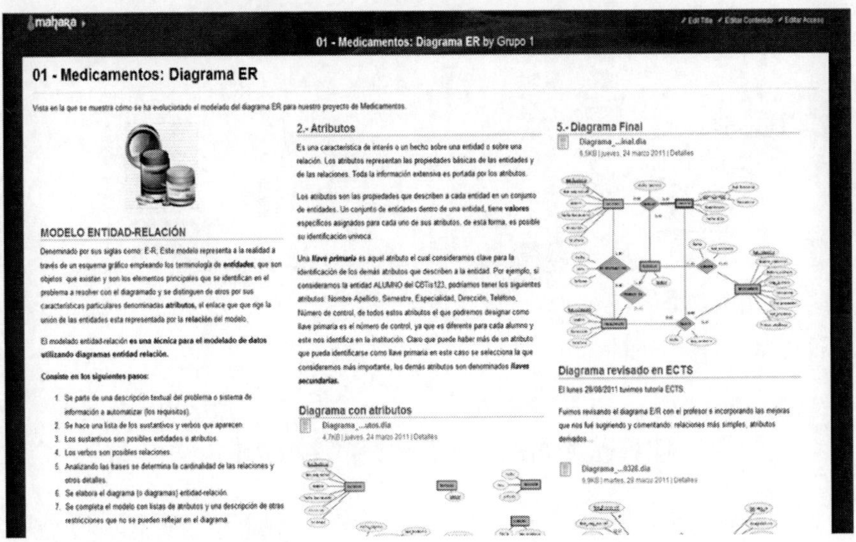

Figure 5. Mahara view example.

Mahara is a web application which offers tools to create and maintain a digital portfolio, as well as different social functionalities which provide interaction among users. It allows you to create versions of the content you want to show (views) as you can see in Figure 5, in our case of the development of the project through artifacts that compose it. One of the advantages is that it allows us to share the views in Moodle, therefore all the work done could be directly assessed in the platform [15].

Mural.ly is a web platform which provides us with the necessary tools to create collaborative walls. It offers different graphic elements and objects as well as the possibility to add web content: photos, videos or links in real time. It includes a chat for users where the comments can be stored so that it can be tracked a record of users' work. Although it is not a Moodle module, the wall can be embedded to Moodle. Moreover, Mural.ly tool also lets you do a presentation from a created wall; it is very helpful to organize work (Figure 6).

In the hypothetical case a challenging methodology was used in the development of the project, Mural.ly could be used as a Kanban board. It is a strategic technique used to control the progress of the work carried out under the context of a production line.

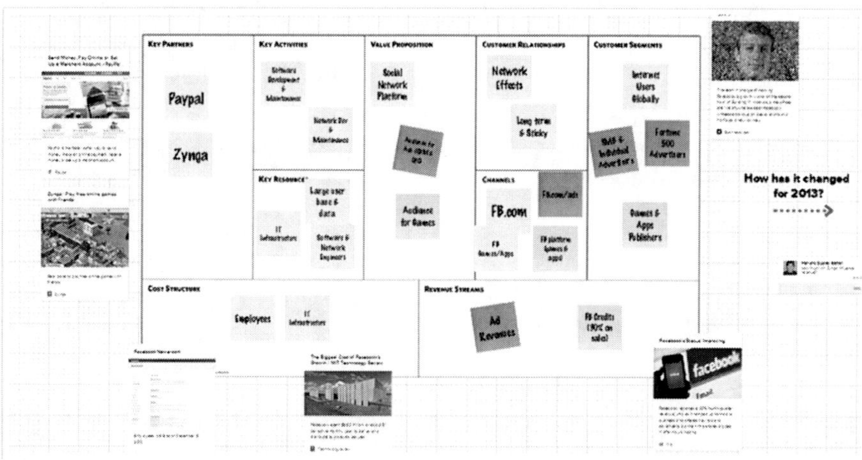

Figure 6. Mural.ly example [16].

PBL ASSESSMENT IN MOODLE

We have just seen the tools which allow developing PBL in Moodle; however, in this section we focus on how to assess the work done by groups of students so that the teacher can grade them and consequently students can be aware of how they have to do the work, what they do right and what they do wrong. Let's look at the assessment tools included in Moodle.

Scales

In Moodle, there is a defined rating scale (0 to 100) used by teachers to assess activities such as forum messages or assignment. But it is possible to define other scales (numeric, Boolean, etc.) to suit a specific assessment. Teachers can start off a simple scale of pass/no pass to more complex scales (eg. poor, just, fair, good, very good, excellent) which can serve to assess the stages of a PBL development.

Outcomes

The tool 'outcomes' helps to establish grading items which can be included in the assignments teachers design, those items could be assessment criteria or competences. Teachers can find track of these items in Moodle gradebook. In PBL we can establish certain competences to be acquired when doing assignments and then do an assessment based on the established competences.

Feedback

Feedback means comments and/or files which the teacher can give to every assignment uploaded by each group of students. The purpose of feedback consists on explaining the weaknesses or strengths of their exercises, so students are aware of how they have completed the assignment and improve it. For example, if a student is developing a project phase, s/he hands in an assignment and a teacher indicates him/her what he/she needs to improve, then the student uploads the assignment again until he completes the assignment properly.

PDF Annotation

Since the upgrade to Moodle 2.6 teachers are allowed to comment directly on students' work if students have uploaded them in PDF format (Figure 7). This is a more direct way to give feedback since teachers can focus, for example, on a paragraph or sentence of each submitted work.

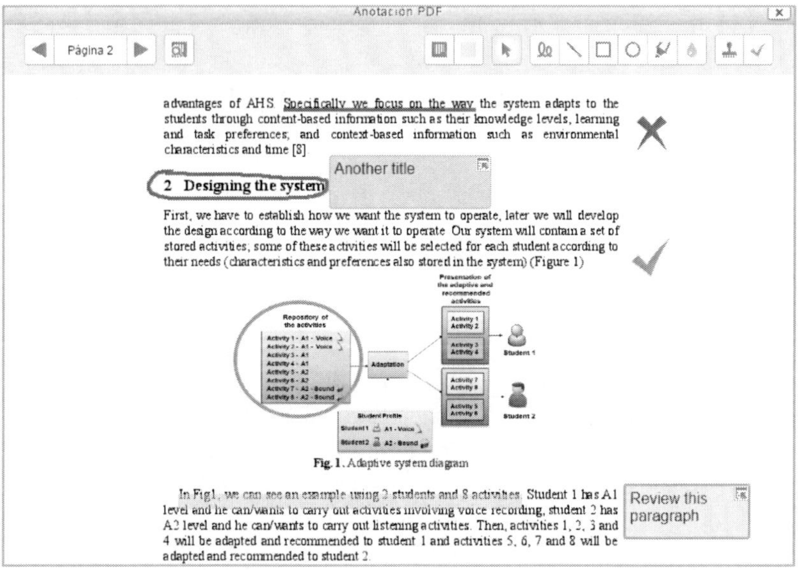

Figure 7. PDF annotation tool [17].

Marking Guide

The marking guides are used to establish assessment criteria for assignments and evaluate them. Thus, teachers can perform more specific assessments, so the grade of each assignment will be based on the established assessment criteria.

Rubrics

Rubrics are advanced grading forms that are useful for ensuring consistent grading practices for criteria-based assessment, rating values and characteristics or objectives corresponding to each aspect-value pair (Figure 8).

In Moodle teachers can configure and use rubrics in the assignments they want. With this tool teachers can assess the work of a group of students more objectively and students can see in their evaluation reports the comments and grades given by the teacher.

Rúbrica

Organigrama	El organigrama es claro para entender el diseño del programa. *2 puntos*	El organigrama no es muy claro, pero se entiende el diseño del programa. *1.5 puntos*	El organigrama está incompleto y es difícil entender el diseño del programa. *0.5 puntos*	No tiene organigrama. *0 puntos*
Resolución	Utiliza las instrucciones y algoritmos más adecuados para resolver el ejercicio. *2 puntos*	Utiliza las instrucciones y algoritmos necesarios para resolver el ejercicio, aunque no son los más adecuados. *1.5 puntos*	Utiliza instrucciones y algoritmos que no son acordes para resolver el ejercicio. *0.5 puntos*	Utiliza instrucciones y algoritmos que no resuelven el ejercicio. *0 puntos*
Funcionamiento	El funcionamiento del ejercicio es completo. *3 puntos*	El funcionamiento del ejercicio no es completo, aunque tiene fallos sin importancia. *2 puntos*	El funcionamiento del ejercicio tiene fallos importantes. *1 puntos*	El ejercicio no funciona. *0 puntos*
Identificación de componentes gráficos y variables	Nombra correctamente todos los componentes gráficos y variables. *1.5 puntos*	Nombra correctamente la mayoría de los componentes gráficos y variables. *1 puntos*	Nombra correctamente sólo algunos de los componentes gráficos y variables. *0.5 puntos*	No nombra correctamente ningún componente gráfico ni variable. *0 puntos*
Diseño gráfico	El diseño gráfico tiene todos los componentes, es claro y estructurado, y además es amigable. *0.75 puntos*	El diseño gráfico tiene todos los componentes y es claro y estructurado. *0.5 puntos*	El diseño gráfico tiene todos los componentes, peno no es claro y estructurado. *0.25 puntos*	Al diseño gráfico le faltan algunos componentes necesarios o utiliza componentes que no son correctos. *0 puntos*
Documentación interna	Aporta una documentación al código para estructurar y/o entender el código claramente. *0.75 puntos*	La documentación aportada es la justa para estructurar y/o entender el código. *0.5 puntos*	Aporta documentación pero es insuficiente para estructurar y/o entender el código. *0.25 puntos*	No aporta documentación. *0 puntos*

Figure 8. Rubric example in Moodle.

DOING PBL IN MOODLE

In this section an example of using PBL in a Database Systems course in Engineering is described. It is divided into six stages following the classical protocol for implementing PBL [18] as we can see in the figure 9.

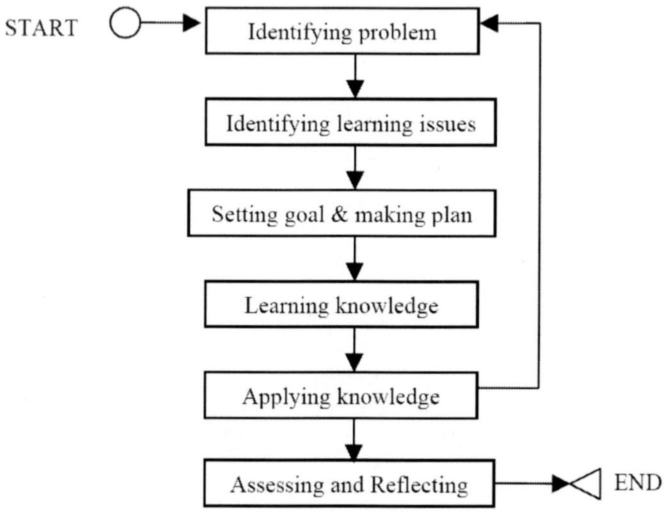

Figure 9. PBL protocol [18].

Database systems is a 6 ECTS credit course, meaning that students must spend 150 hours to complete the course, 65 of them in class activities (lecture, seminar / laboratory and mentoring) and 85 hours of non-attendance which are mainly designed for self-study and doing activities. The main objective of the course is to learn basic concepts related to databases, specifically to learn and develop fundamentals of planning, design, management and maintenance of databases and learn and use programming languages for manipulating databases.

Considering the project to be carried out, students, divided into groups, tackle the definition, the design and the development of a real project of database which strengthens their process of acquiring the competences established in the course and their tracking will be carried out by mentoring. The proposed activity will be done during a four-month period lasting fifteen weeks approximately.

The project consists in the designing and building of a database application and it is finished after the completion of the three phases planned sequentially.

- Design (Entity relationship model, relational diagram)
- Implementation (Converting tables, design BBDD in ACCESS)
- Interaction (Create forms to modify and query database and reporting).

In the following subsections we describe how to do step by step the development of PBL within Moodle (figure 10).

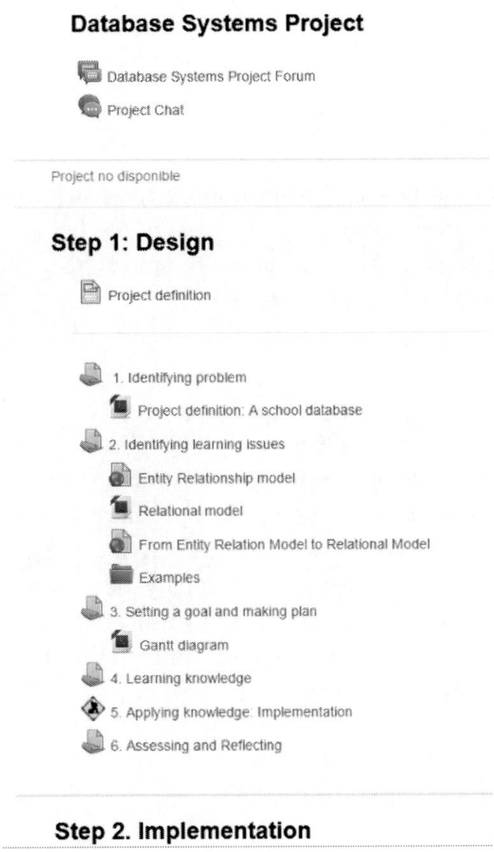

Figure 10. PBL in Moodle example.

Identifying Problem

Moodle tool of grouping is used for the distribution of students in groups. Likewise, each group will be assigned different activities. First, students are provided with a brief description of the project. It would be similar to the memorandum of the project creation in projects management. Then, the group should collect the requirements and develop the project scope heading. This document describes in detail the deliveries of the project and the amount of work necessary to achieve these deliveries. This preliminary phase concludes with the development of the work breakdown structure. Each group of students will upload both documents to the Moodle platform and the teacher will check and approve them before students start with the following phase. The teacher will also upload the final document.

Identifying Learning Issues

Starting from the heading students pose the teachers the concepts they should learn to develop a project and the role of teacher is to provide students with them. In this case, the teacher will upload the information through Moodle resources, files, links, videos, etc.

Setting a Goal and Making Plan

From the material provided by the teacher and the scope heading and the work breakdown structure, the group must define the activities, identify the relations among them (locked or unlocked activities), estimate the resources and the time required to finish them and, finally, develop the project schedule. Ultimately, the group must upload the project's Gantt diagram. These deliveries will be done using Moodle assignment resource, thus, the solution uploaded will be only seen by the teacher.

Learning Knowledge

Once the documentation provided by the teacher and the subsequent students´ production is checked and approved, students must consult and acquire the skills necessary to the completion of the proposed tasks. In this moment the communication plays a vital role for both students and teachers since it is time to ask, clarify doubts and even provide new material if necessary. This is a collaborative process where the groups and the teacher communicate by a single Moodle forum to resolve doubts and problems. At the end of this learning process, the group must create a document pondering over the acquired skills, before passing to another phase the teacher will check it just in case s/he finds lacking of learning.

Applying Knowledge

In this phase we will use the tool TechProject, described in the previous sections, so that the groups can plan the work accordingly. The generated information, that is to say, definition of the project, score heading, tasks and milestone, will be firstly introduced to students. Thereafter the groups manage and execute the project autonomously with the supervision and guidance of the teacher. The phases of the life cycle of the project are:

1. *Database design*. Students, based on the requirements of the proposed information system data, should make the database design. This design includes the conceptual, logical and physical design. The obtained conceptual scheme entity/relationship will be a high-level description of the information contained in the database, it is independent from the Database Management System (DBMS) that will be used. The logical scheme describes the database structure according to DBMS that will be used (relational, object- oriented …). In this case it will be a relational model. Finally, the physical scheme is a description of the database implementation using the data definition language supported by the specific DBMS to be used. For this experience it will be used Microsoft Access. Each group will upload the schemes in a Pdf document to Moodle and the teacher will write the comments down directly in the Pdf.
2. *Implementation*. The groups, after getting the teacher approval to the previous phase delivery, should create the database in Microsoft Access and complete the initial load of data into the tables of the database. The groups will send the database to the teacher so that s/he can assess the implementation and specifically if the load of data to the database considers all the particularities of the approved design.

3. *Interaction with the database.* At this stage, the groups must create the user interface which allows them to access and manipulate the database data. In this case, the user interface will be made up of a set of forms. The desired goal is to ensure the groups maintain the proposed forms. As assistance the teacher offers the group usability guidelines for the design of forms. The students must upload to the platform the database and the relation of SQL queries that underlies in each access form to the data of the database in a PDF document. Within this phase the groups will also write reports as a data summary which must be uploaded to the Moodle platform.

During the project, students will take advantage of all the tools provided by the module TechProject for the tracking and control of the project.

Assessing and Reflecting

The assessment will be conducted using rubrics throughout the whole process, particularly, when finishing each phase, but mainly at the project closure. In the final assessment it will be used both hetero-evaluation by the teacher who will assess the quality of the final project developed through rubrics and comments and co-evaluation where the final project developed by a group is assessed by the rest of the groups using the same tools (rubrics and annotations/comments). We consider it of utmost importance that students know to assess, make decisions and express judgments which affect others effectively. The rubric will be known beforehand by students so that they are conscious of how the evaluation is carried out.

The activity Workshop in Moodle is the most appropriate for allowing others groups to assess other mates' projects. Each group observes how the others solved their projects, thus, enriching in a certain way the insights and learning possibilities. Moreover, the student acquires others cross-curricular skills such as reaching an agreement with the rest of the members of their group and grading the work of the others groups and improving their critical-thinking.

To conclude the assessment, students will be required feedback about their evaluation so that they reflect on their own learning and the development of the work itself.

Conclusion

This paper details the approach of using e-Learning techniques to support PBL based on its functionality to develop critical thinking, to promote self-learning and students' autonomy and the underpinning constructivism on which they both are based. The conceptual exploration, the decision making potential and the analysis control of PBLs are potentialities parallel to such features in online environments. Within our context of engineering, results are measured in terms of the course "database analysis" acquisition in the short term and the real development of competences- planning, design, management and maintenance of databases- which should be materialized in the long run. Thus, the solution of a student-centred and a problem-based approach to the acquisition of practical knowledge in this field allows instructors to apply their role of facilitator and encourage students to explore, analyze and make reasoned decisions by working on quasi-real simulations. Learning is enhanced through the completion of six stages (from the identification of the problem and setting the objectives to how to apply the knowledge and to reflect it on the outcomes). The ongoing results indicate that the adoption of an online instructional constructivist approach can support PBL in delivering exploration of issues arising from students centered scenarios and the development of skills in critical thinking, decision making, team-working and problem solving. As a way of active learning, Problem-Based Learning promotes the construction of new competences and integrates learning with real life dynamics, where learners learn to collaborate, develop knowledge and exchange ideas.

References

[1] D. Brabazon, L. Donovan, M. Melia, M. P. O'Mahony, A. Egan and B. Smyth, 2012, Supporting Problem-based Learning in Moodle using Personalised, Context- specific Learning Episode Generation, *Proceedings of the 1st Moodle Research Conference (MRC2012)*, Crete, Greece, 14 September 2012, pp. 164-171.

[2] R. E. Slavin, "Synthesis of Research on Cooperative Learning," *Educational Leadership,* 38(8), 1981, pp. 655–660.

[3] Mezzari, "The use of Problem-Based Learning (PBL) as reinforcement for students using the moodle learning environment", *Revista Brasileira de Educação Médica,* 35(1), 2011, pp. 114–121.

[4] J. E. Agudo, R. Hernández-Linares, M. Rico, and H. Sánchez, "Generic competences: Competence development in students of the bachelor of engineering in industrial design and product development", *Formación Universitaria,* 6(5), 2013, pp. 39–50.

[5] W. N. Bender, Project-Based Learning: Differentiating Instruction for the 21st Century, Corwin, Thousand Oaks, California, 2012.

[6] M. W. Goodwin, "Cooperative Learning and Social Skills: What Skills To Teach and How To Teach Them", *Intervention in School and Clinic,* 35(1), 1999, pp. 29–33.

[7] Denayer, K. Thaels, J. V. Sloten and R. Gobin, "Teaching a structured approach to the design process for undergraduate engineering students by problem-based education", *European Journal of Engineering Education,* 28(2), 2003, pp. 203–214.

[8] C. Perrenet, P. A. J. Bouhuijs and J. G. M. M.Smits, "The Suitability of Problem-based Learning for Engineering Education: Theory and practice", *Teaching in Higher Education*, 5(3), 2000, pp. 345–358.

[9] J.-N. García, "Problem based learning: An illustration of a psychoeducational application model", *Cultura y Educación*, 14(1), 2002, pp. 65–79.

[10] M. Fernández Martínez, J. N. García Sánchez, A. M. de Caso Fuertes, R. Fidalgo Redondo, and O. Arias Gundín, "El aprendizaje basado en problemas: revisión de estudios empíricos internacionales", *Revista de educación*, 2006(341), 2006, pp. 397–418.

[11] C. Tosun and Y. Taşkesenligil, "Using the MOODLE Learning Management System in Problem Based Learning Method", *International Online Journal of Educational Sciences,* 3(3). 2011, pp. 1020-1045.

[12] P. Sancho, J. Torrente, E. J. Marchiori and B. Fernandez-Manjon, "Enhancing moodle to support problem based learning. The Nucleo experience", *2011 IEEE Global Engineering Education Conference (EDUCON)*, Amman, Jordan, 5 April 2011, pp. 1177–1182.

[13] Moodle plugins directory: Technical Project. https://moodle.org/plugins/view.php?plugin=mod_techproject. Accessed: 18 September 2014.

[14] Moodle plugins directory: PBLTool (CLPMTool). https://moodle.org/plugins/view.php?plugin=block_pbltool. Accessed: 28 September 2014.

[15] PBL with Moodle and Mahara. http://kristinahollis.wordpress.com/2013/03/15/pbl-with-moodle-and-mahara/. Accessed: 10 September 2014.

[16] Mural.ly. https://mural.ly. Accessed: 05 November 2014.

[17] Moodle plugins directory: PDF submission. https://moodle.org/plugins/view.php?plugin=assignsubmission_pdf. Accessed: 05 October 2014.

[18] Y. Miao, S. J. Holst, J. M. Haake and R. Steinmetz, PBL-protocols: Guiding and Controlling Problem Based Learning Processes in Virtual Learning Environments, *International Conference on Learning Sciences*, Michigan, USA, 14 June 2000, pp. 232-237.

In: Project Based Learning on Engineering ISBN: 978-1-63482-217-6
Editors: J. Valenzuela-Valdés et al.

Chapter 7

PROBLEM-BASED E-LEARNING (PBL) USING SYNCHRONOUS VIRTUAL TEACHING

Juan Arias-Masa[*] and Rafael Martín-Espada
Departamento de Ingeniería de Sistemas Informáticos y Telemáticos
Centro Universitario de Mérida, Universidad de Extremadura,
Mérida, Spain

ABSTRACT

The experience described below shows how Problem Based Learning Methodology and Synchronous Virtual Classroom SVC have been combined in the subject Interconnection Systems, in second year course of the Telematics Engineering Degree taught at the University Center of Merida, University of Extremadura. We outline the procedure for distributing students in groups and the way for arranging virtual meetings among PBL groups for developing their own work. Furthermore, students have been encouraged to use at least two SVC technologies and to make a brief comparative analysis about them. In addition, a final student coevaluation (collective assessment) has been made during the presentation.

[*] Emails: juanaria@unex.es

INTRODUCTION

The term Synchronous Virtual Classroom (SVC) appears in the article by Yuste et al. [1], where it is specified that "an SVC is a space for teaching and learning in which computers, with communication networks, make possible that a real class can take place without the physical presence of teachers and students". However, in engineering teaching, one of the main issues continues to be adequate training to construct and deliver high-quality courses [2]. And within these quality criteria, Problem Based Learning (PBL) is one of the most effective methods [3]. Based on this concept, as a tool, SVC has been used in the subject "Interconnection Systems" taught in the second year of the Degree in Telematics Engineering at the University Center of Merida of the University of Extremadura. Also, the class' main focus was to incorporate the PBL methodology during the term 2012-13.

In this chapter we will describe the methodology and the final conclusions obtained in this experience. The entire project was developed throughout the whole half / quarter of the course. The subject belongs to the second-year grade in order to ensure that all the students were aware of the PBL methodology because they had worked these issues in the subject Network Fundamentals during the previous year [4]. Many more examples can be found in ref. [5].

As SVC tools, two of the most frequently used products nowadays have been selected. First, a proprietary software tool from Adobe, Adobe Connect, has been used, which is one of the most widely studied and used in the research community, and, on the other hand, an unrestricted-use software tool as Google Hangout; both meet the basic requirements needed to perform the PBL. In this case, the basic requirements for achieving this goal are consistent with the minimum requirements to be met or which must have an SVC specified in [6] and [7]. This point has allowed the research team to assess not only the use of PBL from SVC but also to compare the performance and ease of using such tools.

PROBLEM TO BE SOLVED

Students were encouraged to use the PBL methodology to work on "Error control at the Data Link level of the OSI model ". Also, different AVS platforms for working in groups were used. As a result, it was necessary to review the content of the common phases involved in the development of the PBL method, described in [8].

METHOD

As noted above, we started from the level of knowledge that students have of PBL Method but not about the use of SVC. However, since they are engineering students and more precisely future Telematics Engineers, training on the use of SVC was minimal, although such training was based on the PBL method. Thus, the first work to be developed by students was to gain deep understanding of both tools that would be used throughout the forthcoming course in order to solve the original problem.

Although the four basic phases included in the PBL methodology remained, namely:

- Phase I. Define the problem.
- Phase II. Research
- Phase III. Making a project
- Phase IV. Presentation of the solution.

Within each phase, a modification was made regarding the operation to which students were accustomed to use or the way they had used the PBL methodology in other degree subjects. Let us discuss how each of these phases has been carried out.

PHASE I. DEFINE THE PROBLEM

This phase takes place in a face-to-face classroom of approximate 50 minutes, based on the traditional methodology of PBL; in this case, the session holds through SVC. In other words, from the beginning and without any prior training, they started using the SVC as a true meeting room for the entire group. For this purpose, a virtual space for the subject of Systems Interconnection is provided. It is used the same virtual area that the University of Extremadura has enabled for it through the URL to access the SVC, as well as a basic manual for logging into the SVC platform. This manual was designed specifically for the students of Systems Interconnection subject for initial access, and the basic steps are listed below:

- Enter the internet address where the SVC is hosted and to which the students want to access. The result of this experience can be seen in Figure 1, where the students (or any Internet user who knows the address) only have to enter his name, though his name and surname recommended, and come into the classroom.
- Run the basic test of audio settings. With this test, the user can detect audio performance both emitting and receiving.
- Identification within the room. In this step, it is recommended that the user of the SVC say hello in the chat. In this way, he will be identified and the room administrator will be able to evaluate the exact operating function of the participants. Within this same step, webcam must be turn on if it is necessary, especially if the participant has one and have been assigned the proper rights to do so.

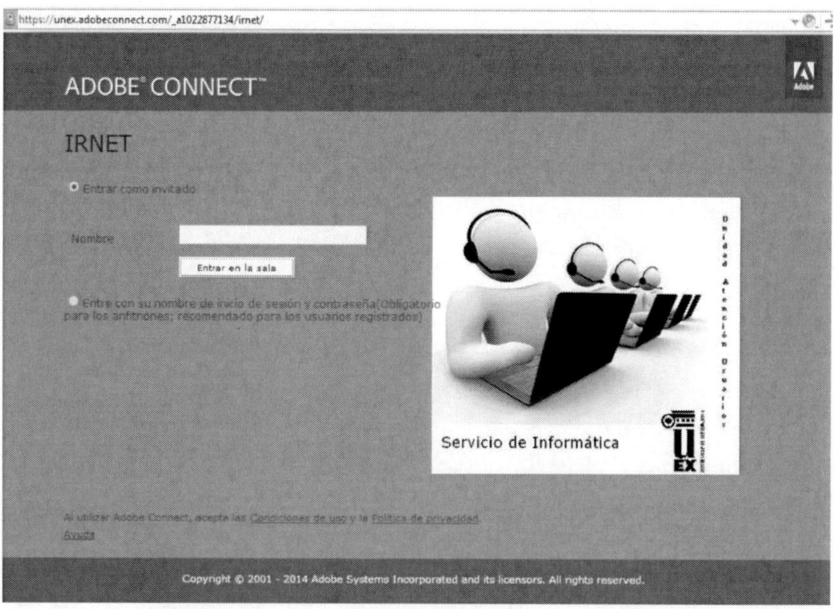

Figure 1. Synchronous Virtual Classroom used for this experience.

In addition to these basic steps, it is helpful a pic of the classroom illustrating the basic elements that an SVC contains. Some of them are described in the paper [9]. Attendees are categorized into three main groups:

participants, presenters and hosts. The meaning and role of each group have "find out" students. And the other element that appears briefly described is chat, indicating that it is always available and it is where attendees must write in whenever there is a communication serious problem.

Step 1. Selecting the Group

The first step is creating groups of students for this initial version. It was easy to do it using AVS, mainly because Adobe Connect provides with the survey functionality. It allows each participant to choose the Group 9 (they do not know who is in that group) (see Figure 2) and then the results can be viewed, and thus completed by the survey. To sum it up, is a random participants in each group choice, so none knows of no other choice and they are also physically located away. The proposal was of 6 groups, given that the total number of students was 30, whereupon it would have 5 students per group.

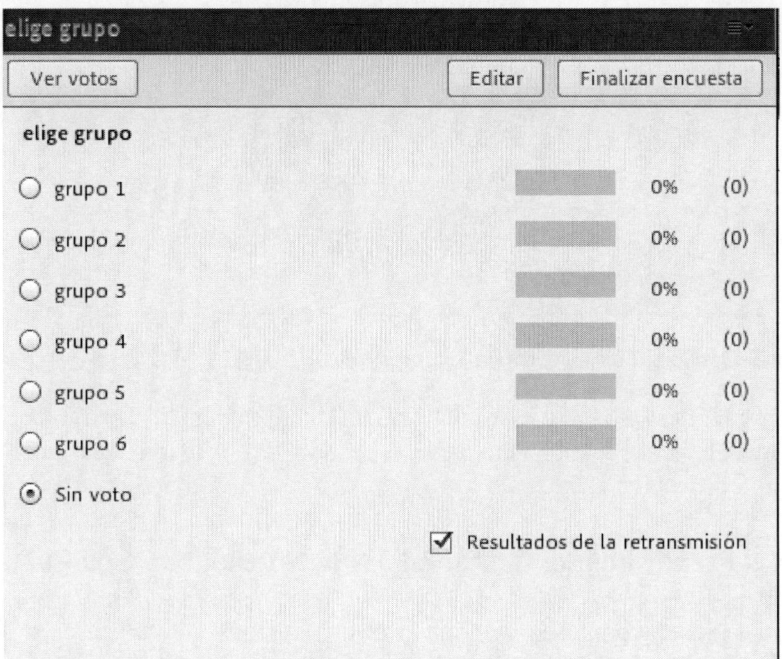

Figure 2. Graphic image on the "pod" Adobe Connect survey.

Randomness will be key in the formation of the group which will make the coordination and follow-up successful. This last statement is based on data obtained from the final poll. See in the Figure 3 the question “do you think that members of the group comply with the established rules better when the group is formed ramdomly?” In the figure 3, we can see that a very large majority if we add the seventy-five per cent more 20 that also agree, concludes with a 95% which is almost unanimity, especially if the other 5% is missing that has been in the section considering it indifferent.

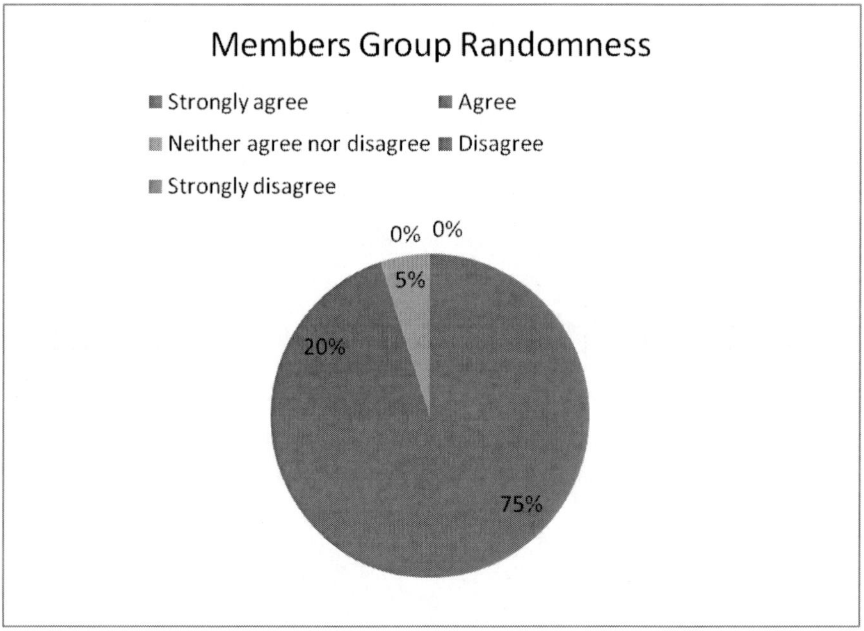

Figure 3. Opinion of the students on the formation of groups.

During this phase, the formulation of the problem to be resolved is presented. It can be found in the previous section “Problem to be resolved”.

Step 2. Preparation of the Initial Team Notebook to Collect Data

Some groups proposed working with a “notebook” to collect data. This data are mainly the rules of operation, as well as the distribution of tasks or responsibilities. To carry out this, each group meets in a room of Adobe Connect, because this AVS allows working groups and each group has its own meeting room and can she have available all the resources needed to carry out such a meeting. This collecting data "notebook" (agreed by common consent by each and every one of the components of each group) gets uploaded into the virtual classroom under the subject of "workshop activity". That allows collaboration of students with their peers, in addition to the evaluation of the teaching staff of the subject itself.

An important section in the notebook computer is the categorization of responsibilities of the group. The teaching staff always offers at least four roles, namely:

- Coordinator (permanent)
- Secretary of meetings (permanent)
- Responsible for dividing the group work (rotates)
- Responsible for drafting documents (rotates)

This initial proposal was accepted by the majority of the groups, but with slight modifications. Thus, the most significant changes were made by y the role of Coordinator and Secretary of meetings. Last, they agreed that the roles have to be rotating and also, they set the dates according to the weeks where role changes are made.

Finally, another very important paragraph of the "notebook" (to collect data) is the "Operation rules", and that the teaching team proposes the following:

- Meetings cannot be missed unless the absentee has a justification.
- Everyone has to be responsible for their own work.
- Have predisposition in helping the other students if they have not finished the task.
- Punctuality when it comes to attend the meetings of the group.
- If a person violates any of these rules three times, instead of taking only a task, will get you one more.
- If you violate these rules with a maximum of six times, will penalize with the immediate expulsion from the group.

With these rules proposed by the teaching team, each group drew up their own rules and evaluation of them. We can conclude that they were much stricter in their wording and as we'll see in the results, especially in its execution, which makes them good for acquisition of skills expected it to get this work. As an example of the mentioned, the following rules of any group are:

- No member may leave the meeting before the end.
- If any member does not carry out its work, s/he will be placed as a serious lack.
- A third serious offense will result in reject from the group.
- If a member is absent without justified reason to the meeting, s/he will be punished.
- Tardiness at meetings will involve the payment of 50 cents.

Step 3. Initial Agenda

On the agenda, it is important to differentiate between individual group meetings and general meetings.

The first are organized by each of the working groups, and they who have to manage such meetings. Therefore it is possible to make the changes that each group decides.

General meetings are organized by the teaching staff and the assistance of all members of each group.

All meetings, both the specific and the general should be made on the AVS. But while the former are managed by the teaching staff, the latter must be organized, particularly, for each group; and in addition, it is necessary a coordination between the groups to not coincide in time in the AVS. Therefore, here comes a new PBL, in which the management of the AVS is made by any member of the group. In fact, all must study and learn the operation of the AVS as hosts for the Adobe Connect classrooms, and inter-coordination between the groups is also necessary so that they can use the same AVS. This last makes that to make future changes in the initial agenda is difficult.

Regarding the choice of the particular use of the AVS, it is a free choice by the members of the group. Adobe Connect chat and survey functionality will be used for this purpose.

Chat is used by the members of each group to reach agreement among them which is a common available source. In the case of non-changeable schedule conflict, teaching staff will be which alternately chose the time schedule, by placing groups on alternate weeks.

Secretaries of Group Meetings

As a new element concerning the methodology of PBL, the concept of group meeting Secretary arises. This figure will be always occupied by a student of a different PBL Group on a rotating basis. All group meetings will take place every 15 or 21 days. The particular student of the PBL Group's name will be proposed by the spokesman of the group to the teaching staff.

This Secretary will be the note keeper of all the final agreements during sessions. All this documentation is stored in the virtual classroom course where students have access.

Summary of Phase I

This is a summary of all the components of the "notebook" (collection of data):

- Components or members of the group.
- Categorization of responsibilities of each member.
- Rules of the internal functioning of the group.
- Schedule of planned meetings.

Phase II. Research

At this stage it is important to make a distinction between the research sessions from the rest of the sessions. AVS is used as a physical support of the meetings; therefore, it is mostly necessary that the teaching staff attends this session.

First Session of Research

This phase begins in a second session of classes in the AVS of Adobe Connect for all students; each from its Internet location. This session's host is the professor of the AVS room.

The first step is devoted to allow students investigate the working functionality of an Adobe Connect room. Consequently, groups are going to their own Adobe rooms during the first part of the first session. Then, everyone meets at the central room. Each group has a report that will share with the rest of the course participants and groups. The report is the conclusion of the research using the PBL methodology on the use and management of the AVS.

As an initial premise, the teaching staff has students to conclude a summary of the management of the Adobe Connect room as each group of PBL has to hold their own meetings. Therefore, it is necessary that one of the members of the group "opens the room" as host and then make it accessible to other members as participants. Later on, he upgrades them to the role of presenters[1]. In this experience the 6 groups quickly identify the need for roles and quickly learn the basic operation of the room.

The second step of this session is a group meeting for all students. There will be a spokesman for each group to report about its findings. They will conclude with a final report of this meeting prepared by the General-Secretary of the class:

- It is important that a member of the PBL group is responsible for administering and managing the room with the role of a host. However, there should be also a person responsible in case the host cannot attend the meeting. That person will be the back up host.
- All other members of the PBL Group must have the role of presenters in the room to be able to interact and work properly in the Adobe classroom
- Google Talk is established as an alternative mechanism or as complement to the use of the AVS

After students have completed the basic knowledge of the platform, the next step will be how to focus on resolving: "Control of errors at the level of the ISO OSI model link". For this purpose, the students will meet in PBL Groups using the AVS for the meetings. This phase will take about two weeks. At the end, students must deliver a document with a summary of all the data collected.

During the research phase the possibility to use an alternative free virtual classroom is discuss. The team ended up proposing Google Hangout as an addition of Adobe Connect. An objective criterion of evaluation of the two

[1] The three roles that an Adobe Connect classroom establishes are (in decreasing order of permissions): Host, presenter and participant.

platforms (Adobe Connect and Google Hangout) is established and the team agrees to have at least two meetings.

This will allow us to make a parallel investigation on the quality of the AVS boards at the end of the PBL work. Some of the results are shown in the corresponding section.

During this phase, the delivery of results from the groups will not be necessary. At this point the interaction and discussion among the group members is important. They will also be reviewing the process and maybe redefining some tasks.

PHASE III. PROJECT EXECUTION

This is the main phase of the PBL project, the longest one, specifically, it lasts six weeks and during this time at least 2 group meetings will be held in order to be able to raise doubts to the teaching staff and, above all, to achieve that the same teaching team review the situation of each PBL group. Both group meetings have the same working methodology: first of all, students are gathered in the SVC and then, a sub-room is created for each PBL groups. In figure 4, we can see an example of the SVC of an Adobe Connect meeting.

Figure 4. Group meeting example between teaching staff and a PBL group.

Therefore, students can take advantage of a working session as an additional session to those that they need to carry out the work; the only thing to consider is that they must dedicate a short time to speak to the teaching staff, raising their doubts and answering to every issue the teaching staff can expose, in order to get a clear idea of the specific situation of each PBL group. As clear milestones that the teaching team wants to detect through these meetings the following ones can be raised up:

- Cohesion of the group.
- Record and compare the degree of closeness and friendship beyond professional relationships that links the members of the PBL group each other.
- If any weaknesses are detected in the operating function of the PBL group, this is the exact moment for trying to redirect this situation.
- Detect what percentage of the final project is already done.
- A recommendation for each group can be made if there is a significant delay, they are in time and in the proper way or otherwise, they are dealing with some time in advance, though it is not usual.
- Review the equipment notebooks
- It is required that computer notebooks are updated in real time while the groups are functioning and not at the end of the work as if they were one more document to fill.

From the previous work, relevant annotations are written down by teaching staff in order to make a comparison in the second group meeting and check if the group has progressed properly. In addition, these annotations will be contrasted with the final evaluation of students' notebooks.

The output of this development phase or project implementation phase is the formal documentation which students must elaborate and then deliver in the SVC; these documents will be co-rated by their colleagues.

In this particular, PBL requires them to give the following documents:

- Final team notebook. It has to be a complete notebook with exact detail of all the important information extracted from the meetings.
- Document elaborated in the usual framework filled in the developed works: cover, index, summary, content, personal assessment, and bibliography.

- Document for the presentation or defense of the entire work. Usually, it is a presentation in "power point", "prezzi" or any other similar tool with slides or images to be used for the presentation.

PHASE IV. PRESENTATION OF THE SOLUTION

Final solutions to the initial problem must be exposed [10] in the SVC of Adobe Connect by all groups, following the next performance rules:

- All works must be delivered in the virtual classroom of the subject previously to his "defense" in the SVC.
- As there are 6 groups to expose, two final sessions of 50 minutes are reserved so that three groups can participate in each session. A time period lower than one hour and a half facing the computer screen should be spent, as it is very damaging for people and it provokes a significant loss of attention.
- The order of PBL groups' presentations is varying and depends on a draw made by the teaching team during the presentation session of the results. Using a random order can make sure that the order does not influence the co-evaluation that students make and, which is more important, in this way student groups must have their presentations well-prepared, avoiding last minute changes based on strength and weaknesses found in other groups presentations.
- For every PBL group, the presentation order is fixed by the teaching staff at intervals of 5 minutes, and in vivo, as only the teaching team knows that presentation order from the previous draw. Thus, the teaching staff will be giving the floor to the appropriate student in order to make his presentation in the proper order. In this way, it is sure that all students know perfectly the whole final report which is being presented, because they do not know which part they have to present. This methodology is being used by this teaching team for several academic courses for all group presentations, always looking for the goal that the final work is been developed collectively and, above all, all group members gain a deep knowledge of the final report.
- Each PBL group has a maximum presentation time of 20 minutes, which has not to be necessarily exhausted. Once the presentation is completed, there are always 5 minutes so that any other students or

teaching staff can interact with the group about any information appeared in the presentation.

- After the presentation of every PBL, students spend 5 minutes for coevaluating the presentation made by the student group which has just presented its part of work.
- The student coevaluation is made through the application of the scoring rubric shown before, which finally produces an entire value rank between 0 to 100; it follows the rule of discarding those qualifications 25 points far from the mean value and if so, recalculating such mean value. Besides, as it is a well known mean value and the students know that it is the way it is going to be qualified, they are more reflexive at the time of selecting the elements of the score rubric used for evaluating the works. This rule has been applied in previous works of the teaching team and students have been questioned about the readiness of the rule itself, and it can be seen in figure 5 how students think that this method let them be much more reflexive in evaluating the works.

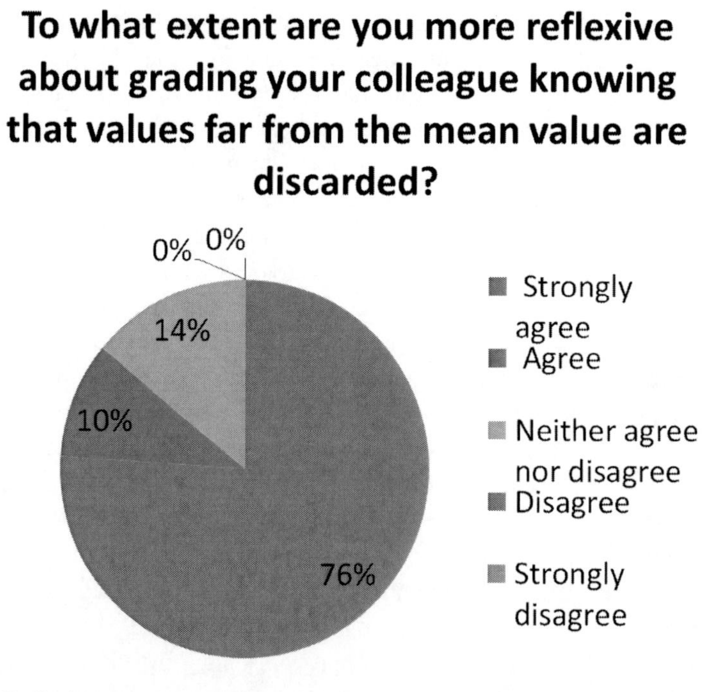

Figure 5. Reflexive value on grading works.

Experience Evaluation

As an PBL evaluation experience, the following elements or parameters have been used:

- Online questionnaire for students' opinion on their own developed work.
- Test of level certification of knowledge about the work topic, which has given very positive results with a distribution of qualifications like that shown in Figure 5. Such test is previously made by the teaching staff on the contents to study.
- Students' coevaluation results about their colleagues presentations.
- Documentation submitted to students review by teaching staff through applying the proper evaluation rubric.

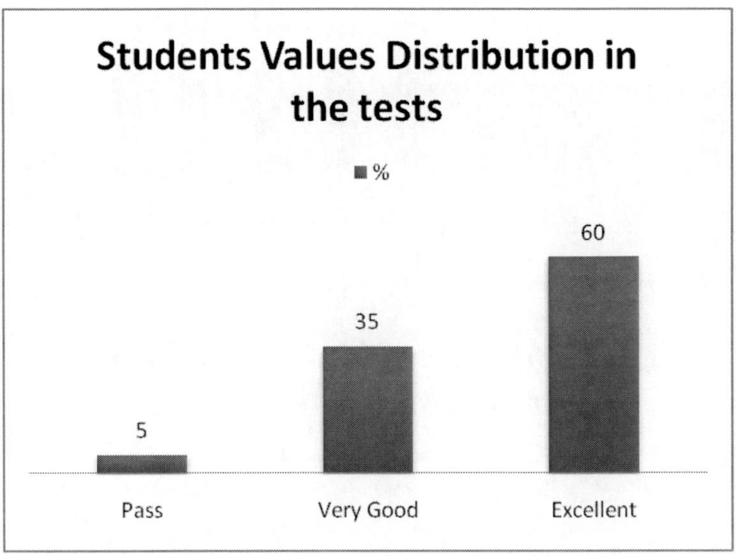

Figure 6. Final Student Values Distribution in the certification test.

From the data registered on the experience evaluation, it is valuable to show in this paper those considered more relevant, and, beyond that, the co-evaluations held by students during the other groups' presentation of results, considering that the group under evaluation cannot participate in the evaluation itself. Therefore, only the co-evaluations of all remaining students and the teaching staff, who always participates as one more element, are considered to work out the mean value. In Figure 7, where the final results of evaluation made by the students is shown, the level of presentation quality rated by each group can be pointed out; the score is restricted from 1 to 10, and it can be proved that most scores are around the mean 8 and below the pass level only a 2% are located.

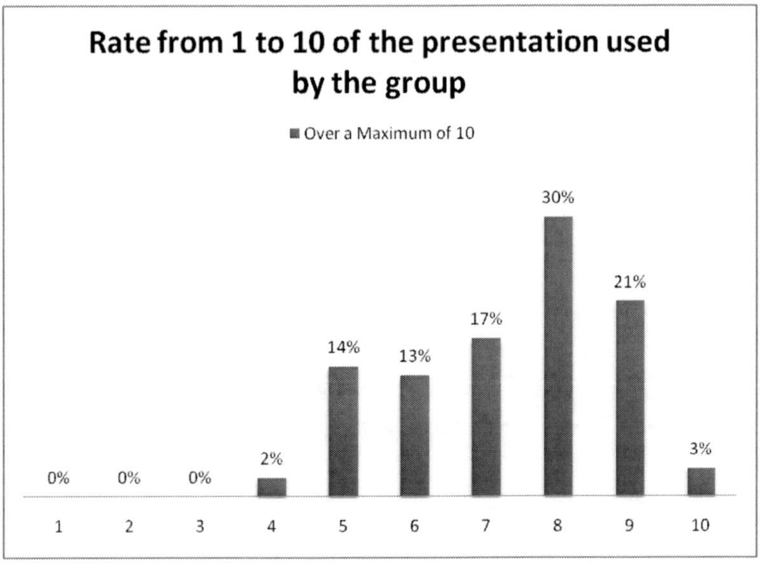

Figure 7. Quality of the presentations used by groups.

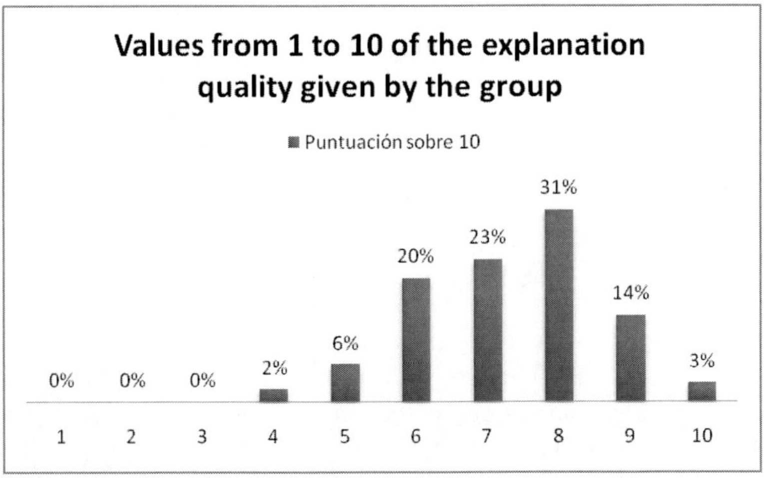

Figure 8. Explanation Evaluation.

In Figure 8 the evaluation about issues explanation can be seen. It is shown that most rates are located between 7 and 8; If both values provide a very good rate and represent 50% of the entire qualifications, the result can be considered very positive.

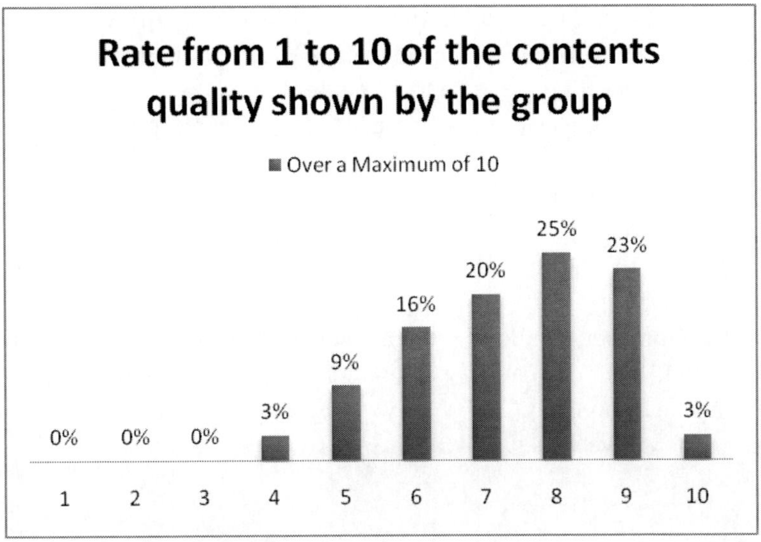

Figure 9. Contents evaluation.

In Figure 9, it can be seen how most values remain around the mean 8 and 9, which is an excellent rate, and this point verifies and coincides with the certification exam passed by the students whose results can be seen in Figure 6; thus, values between 8 and 9 provide more than 50% and these results are highly positive.

RESULTS

One of the SVC accomplished evaluations consist of the comparison between Adobe Connect and Google Hangouts, though the comparative analysis of these SVC is far from the scope of this paper, but an important conclusion is that Google Hangouts is perfectly valid for every individual PBL group meeting but it is not so valid for group meetings. In any case, the percentage of these two platforms usage by PBL groups indicates that 65% of the students used the Adobe Connect room compared with 35% who used the

Hall Google Hangouts, according to the data obtained from the final team papers provided by every group at the end of the activity. Anyway, the following advantages of using SVC platforms over not using them based on other PBL made by these same students are detailed below:

- Capacity to organize a meeting in short time, and, above all, to access anywhere, whereupon the advantage to manage time compatibility between meeting participants is maximum.
- Sense of attendance as close to human perception as physical attendance.
- Each student has to fully play or execute its own role in the group; a passive behavior in the meetings is not possible.
- The new teamwork protocol has allowed them to share information quickly.
- Desktop sharing allows what anybody writes to be seen instantly by all, and everyone can give an opinion about it at the same time.
- In conclusion, SVC is a very positive working tool to make own collaborative works.

CONCLUSION

The experience described in this chapter has been highly successful for both students and the subject teaching staff.

On the one hand, all the students have successfully achieved the objectives established specifically for this subject, also acquiring specific competences, which was the main goal of this research work. In addition to such specific competences, they have been working to develop a set of cross-disciplinary competences that they must acquire throughout the whole degree and in every subject.

On the other hand, the problem-based learning experience has been made on synchronous virtual classroom SVC which nowadays represents a very useful tool for face-to-face, distance teaching and group meetings; as it has been shown through the results in this paper, the experience has been very satisfactory.

Finally, we recommend the use of problem-based learning for the study of any topic of the subject and also the use of the synchronous virtual classroom SVC both for the development of group works as for the final presentation of such work.

REFERENCES

[1] Yuste, R., Alonso, L., & Blázquez, F. (1988). *Synchronous Virtual Environments for e-Assessment in Higher Education. Scientific Journal of Media Education,* 159.

[2] Palloff, R. M., & Pratt, K. (2013). Lessons from the virtual classroom: The realities of online teaching. John Wiley & Sons.

[3] Santos, S. C. D., & Soares, F. S. (2013, May). Authentic assessment in software engineering education based on PBL principles: a case study in the telecom market. In *Proceedings of the 2013 International Conference on Software Engineering* (pp. 1055-1062). IEEE Press.

[4] Miguel Angel Martín Tardio; Juan Arias Masa; Javier Carmona Murillo; Juan Carlos González Macías; Juan Francisco Valenzuela Valdés; Pedro José Pardo Fernández. *Conexión y verificación del cableado horizontal y tomas de red (UTP Cat 5e/6) en un Sistema de Cableado Estructurado.* XXI Jornadas Universitarias de Tecnología Educativa. Valladolid, 2013.

[5] Duncan, M. J., Smith, M., & Cook, K. (2013). *Implementing online problem based learning (PBL) in postgraduates new to both online learning and PBL: An example from strength and conditioning. Journal of Hospitality, Leisure, Sport & Tourism Education*, 12(1), 79-84.

[6] Tan, J. (2013). An Analysis of User Requirements for Virtual Classroom/Collaboration Software. *International Journal of Information Technology & Computer Science*, 9(2), 12-22.

[7] Lajoie, S. P., Hmelo-Silver, C. E., Wiseman, J. G., Chan, L. K., Lu, J., Khurana, C., & Kazemitabar, M. (2014). Using online digital tools and video to support international problem-based learning. *Interdisciplinary Journal of Problem-based Learning*, 8(2), 6.

[8] Miao, Y., Samaka, M., & Impagliazzo, J. (2013, August). Facilitating teachers in developing online PBL courses. In Teaching, Assessment and Learning for Engineering (TALE), 2013 *IEEE International Conference* on (pp. 454-459). *IEEE.*

[9] Ismail, J. (2001). The design of an e-learning system: Beyond the hype. *The Internet and Higher Education*, 4(3), 329-336..

[10] Alonso, L., Gutiérrez, P., Yuste, R., Arias, J., Cubo, S. y Dos Reis, A. (2014). Usos de aulas virtuales síncronas en educación. *Pixel-Bits, Revista de Medios y Educación,* 45, 203-215.

In: Project Based Learning on Engineering ISBN: 978-1-63482-217-6
Editors: J. Valenzuela-Valdés et al.

Chapter 8

PBL WITH M-LEARNING

***Juan F. Valenzuela-Valdés*[1,*],**
***Pablo Padilla de la Torre*[2]**
***and José Luis Padilla de la Torre*[3]**
[1]Department of Computing and Telematics System Engineering, Universidad de Extremadura, Centro Universitario de Mérida, Mérida, Spain
[2]Signal Theory, Telematics and Communications Department, University of Granada - CITIC, Spain
[3]Nanolab, EcolePolytechniqueFédérale de Lausanne, Switzerland

ABSTRACT

This chapter is centered in the innovations of mobile learning (m-learning) and ubiquitous learning (u-learning) related to problem based learning (PBL). First, a literature review of works related to m-learning and u-learning is done. Then, it is presented one m-learning experience implemented in a telecommunication degree regarding the radio navigation course. Finally, it is shown an innovative experience on u-learning. This u-learning experience goes further in the use of mobile devices as a measurement tool for students, providing a great level of practical contact and interaction with the handset, with benefits in the meaningful learning. In addition, the device capabilities let generate a

[*] Corresponding author: J. F. Valenzuela-Valdés. E-mail: juanvalenzuela@unex.es.

"ubiquitous learning environment" to interact with real world and learn the main concepts regarding wireless networks.

1. INTRODUCTION

In recent years, the Information and Communication Technologies (ICT) have been developing in an unstoppable way. In fact, nowadays, we are in what is called the technological revolution of wireless communications, a similar revolution that happened at the time of electricity, television, computer systems or wired communications, which accounted for new business models.

One of the main advantages of this technology is mobility. The facts of not having the access point into the communications network limited to a fixed location and having the transmitting propagation channel always ready favor its expansion, which is faster than any other type of technology. There is a corroborating example: in just five years of existence, there have been more mobile phone users than landline telephone ones. The internet has also taken advantage of this technology, going through what is known as mobile Internet, which allows mobile devices and people to connect to the network anywhere and at any time. This fact has facilitated the emergence of new services and applications on these devices. A group of these new "services" are related to Education and imply the use of wireless devices with learning purposes. Thus, this new technology is also of use in university teaching methodologies.

These new technologies have also been introduced in the Project Based Learning (PBL) methodology. When new portable devices are used in order to develop learning activities, this methodology that uses portable devices (such as mobile phones, PDAs, tablets, Pocket PC, iPod) is called m-learning. Modern learners have increased their needs of learning time and are often forced to study during their lunch breaks, in the evenings and at weekends; and to study at any time they can do it, for instance in their displacements by bus, train, etc. The use of portable technologies makes it easier for learners to study everywhere and at any time they want, being simple for them to transport their learning materials. M-learning also facilitates the "just-in-time" learning, as learners can take advantage of any unexpected available free time since they have their devices with them most of the time.

In the last years, there have been many recent studies on m-learning [1-10]. A first group of studies analyze different ways to use mobile technology in learning [1-6], and also comparisons between e-learning and m-learning [4], pointing out positive outcomes of m-learning when compared. In [5], it is analyzed how the use of mobile technology enables the informal or semi-informal learning, meanwhile in [6] it is analyzed the impact of m-learning in students and teachers. Other works are focused on specific characteristics such as size and weight of mobile devices, as in [7]. In [8], it is analyzed the effectiveness of using different devices to access information with eye tracking. In [9], it is studied the usefulness of a text summarization tool for mobile devices. Finally, one quite complete review, analysis and synthesis of 164 studies of m-learning in the last years is performed in [10], being a reference document.

As it is explained in [4] the concept of "context-aware ubiquitous learning" was introduced due to the increase of different capabilities of the mobile device, such as: geospatial technologies, social networking and different uses of the handset camera (i.e. visual search or image capture).As it is defined in [11], “in a context-aware ubiquitous learning environment, single students are guided to learn in a real world situation with support or instructions from a computer system or using a mobile device to access the digital content via wireless communications”. More studies can be found in [12-13].

In the following section, it is presented two experiences developed by authors on m-learning and u-learning.

2. M-LEARNING EXPERIENCE “TRAINING WITH RADIO NAVIGATION SYSTEMS”

2.1. Introduction to Radio Navigation

Radio navigation is the science of geo-positioning using electromagnetic waves. The electromagnetic waves can be used by a receiver to determine the position in relation to an emitting source. Combining different sets of line of position (LOP), positioning can be obtained. Radio navigation systems can use azimuthal, telemetric or hyperbolic LOPs to obtain positioning. The arising of radio navigation was initiated in the twentieth century and was developed parallel to the development of electronics and radio frequency and microwaves technology.

This Radio navigation PBL m-learning experience is focused on senior-level undergraduate and electronics, telecommunications or naval engineering graduate students and the knowledge of electromagnetic wave propagation is applied. Although there is no prerequisite for the Radio navigation Laboratory course, students should be familiar with the basic theoretical concepts related to wave propagation at radio and microwave frequencies and antennas. Most of these concepts are taught in earlier stages of a telecommunication engineering degree.

2.2. Pedagogical Issues

The cognitive Learning Outcomes (LOs) have been achieved through the combination of theoretical lessons and m-learning experiences in the laboratory, while the psychomotor LOs have been achieved with m-learning experiences. The assessing of the LOs is verified through a test exam that evaluates the cognitive LOs and the evaluation of post-report experiences. The LOs of this course are shown in Table I.

Table I. Lab Experiments, Used Tools and Learning Outcomes

Utilized tools	Learning outcomes (LOs)	
	Cognitive LOs	Psychomotor LOs
Training on directional and telemetric radio navigation system	Knowing the VOR radio navigation systems	Know how to use of VOR in simulator
	Knowing the DME radio navigation systems	Know how to use of DME in simulator
	Knowing the VORTAC radio navigation systems	Know how to use of VORTAC in simulator

The purpose of this experiment is to become familiar with the Very high frequency Omni Range (VOR) and the Distance Measuring Equipment (DME). Many times, we understand the system after receiving instructions of how it works and through its use, which is what we do in practice. Practical understanding of the VOR system will allow a better understanding of what it is explained in the theoretical lessons. The Omni Bearing Selector (OBS) was named in the early VOR days and may be misleading if taken literally. In this practice, it is necessary to become familiar with the use of the OBS selector, as well as the indicator "TO/FROM". The application used is "Nav Trainer Basic" [14] and "Nav Trainer Pro for Pilots" [15], both from "Fleminger media" and available in the Google play store. In these applications the students practice with: 1. Dual Navaids for x-bearing, 2. DME visualization, 3.

ILS VOR and 4. ADF. These concepts are introduced previously in theoretical lessons. Figure 1 shows a screenshot of Nav Trainer application.

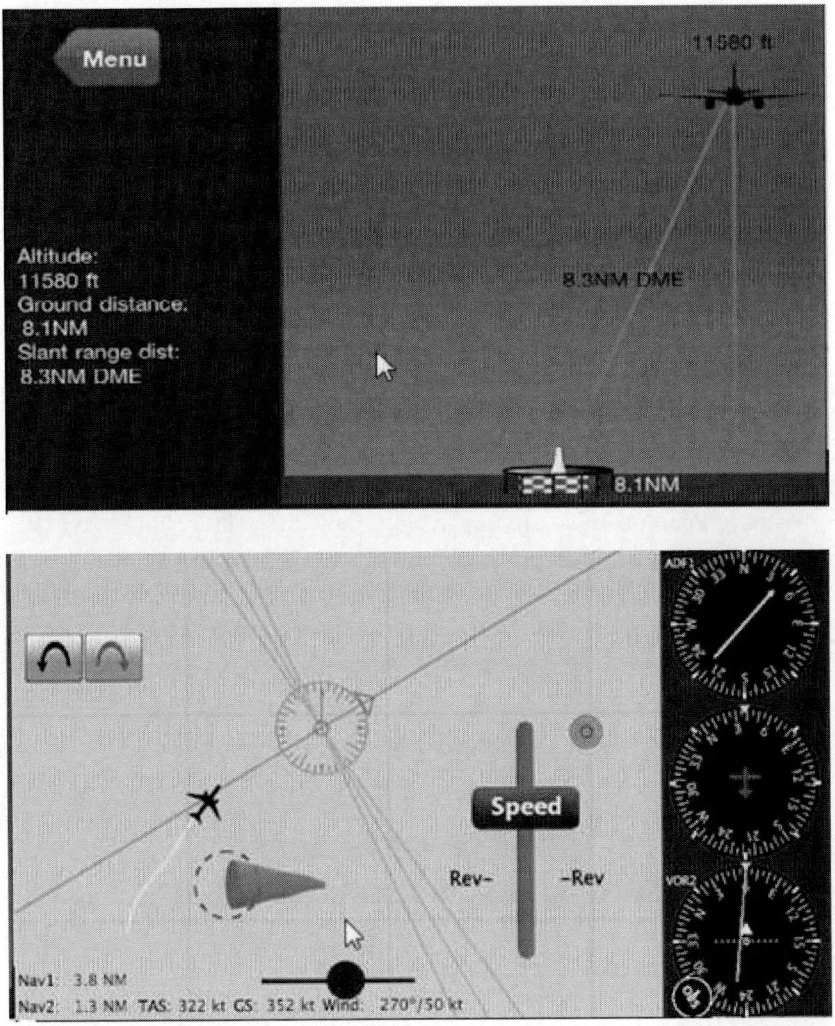

Figure 1. Screenshot of an application for m-learning VOR Simulator.

In this experience, the students carry out a demonstration to the teacher of the use of the different simulation software's and the practice is evaluated with this information. Therefore, in this experience the students must not deliver a post-experience report.

2.3. Results

The Psychomotor learning outcomes have been evaluated through a post-experience report. Figure 2 shows mean and standard deviation scores in the test exam and post-experience report for all the academic years. Figure 3 shows mean test scores and the mean report scores divided into academic years.

From these figures it is possible to extract the following conclusions:

- Scores in the post-experiences reports are higher than scores in the test exam.
- The mean scores in reports are similar in the different years. However, there is a higher variance in the mean scores in the test.
- Scores in the post-experiences reports are higher than scores in the test exam. Three different factors can cause this finding: first, with the instructional method used, the students are more motivated in the laboratory than in the theoretical course, second, the cognitive and the psychomotor LOs are detached, thus different results have been obtained, and third, probably the students in a technical career tend to develop more psychomotor abilities. That is, in the course conditions psychomotor LOs seem to be easier to be absorbed than cognitive LOs.
- The mean scores in the reports are similar during the different years. However, there is a higher variance in the mean scores in the test.

3. U-Learning Experience "Training in Wireless Networks"

3.1. Introduction to Wireless Networks

In a broad and general sense, we understand Wireless Communications as the exchange of information using the electromagnetic spectrum. Cellular technology has emerged in the evolution process of the communication systems, accelerated by the high demand for mobile services, data processing and other complementary-based cellular network services. Nowadays, a variety of different standards and networks coexist in the mobile telephone world. The cellular systems that continue working are the second generation

2G (consisting of digital systems: AMPS, GSM, CDMA, PDC with average speeds of 9,600 bps),the 2.5 generation2.5G (with GPRS, CDMA rev B with speeds 64Kb/s), the third generation 3G (WCDMA and CDMA2000 operating systems technologies, with speeds from 144Kb/s to 2Mb/s), and the new the 4G systems, which offer high-speed data transmission from 2Mb/s up to 1 Gb/s. The objective of this experience is to validate that the students learn to discriminate and identify the different networks and different network capabilities.

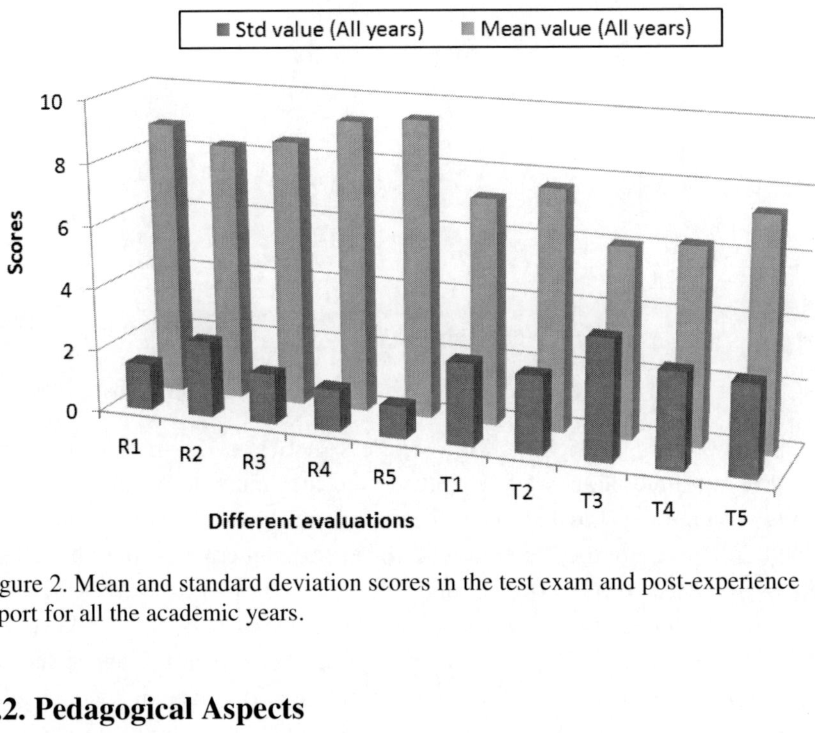

Figure 2. Mean and standard deviation scores in the test exam and post-experience report for all the academic years.

3.2. Pedagogical Aspects

For each of the two experiments that comprise the laboratory-course, a theoretical section (with sessions of 2 hours) and detailed instructions of the tasks to be done are provided, along with deliverables (slides of both parts are available online and a discussion forum is enabled in order to solve questions and doubts). The laboratory sessions are based on a ubiquitous learning methodology in which students can conduct the experiment at any time and any place, but the teachers have a schedule during which they are available to solve doubts face to face. The experiments are designed to be carried out in groups of two or three students. The students are required to deliver a report of their work for each experiment within a fixed timetable. At the end of the term, a final mandatory exam evaluates their learning outcomes.

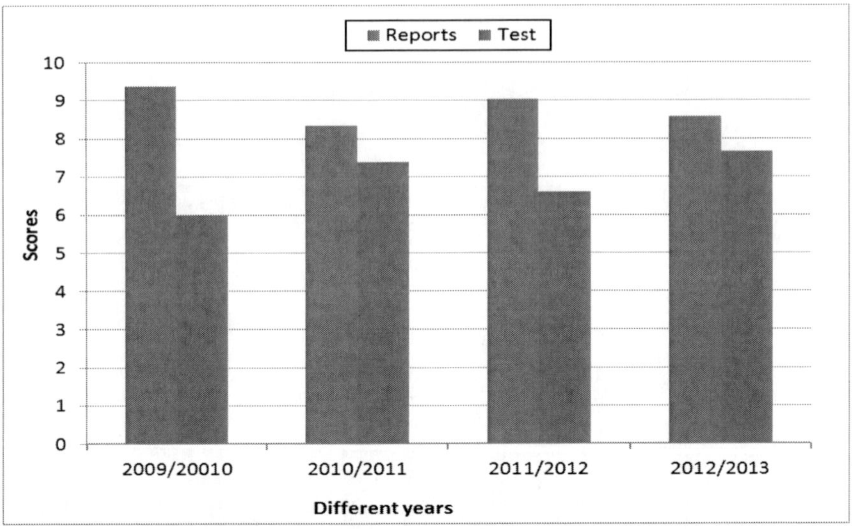

Figure 3. Test scores and the mean report scores divided into academic years.

The laboratory, designed with wireless networks in mind, must be developed in a ubiquitous way. Therefore, the tool must be installable in the students' computer or device. The license of the required software must be available for free to the maximum number of students as possible. The laboratory-course covers two different aspects of wireless networks: the first covers the planning and dimensioning of wireless networks, related to experiments 1, 2 and 3. The second implements experiments 4 and 5 and is related to the measurement of the physical propagation layer in a novel way through mobile applications. Laboratory experiments, used tools and learning outcomes pursued with the completion of these projects are shown in Table II. Table III shows skills to be acquired and level of skills development (1 minimum - 5 maximum).

To complete the project, students have to perform a report. They also had to defend the report and each of the decisions made throughout the project, in class to their peers.

In this project, the PBL methodology combined with the methodology of "m-learning" was carried out through mobile smart phones. Therefore students were free to carry out the project when and where they wanted. This is one of the advantages of the m-learning.

Table II. Laboratory Experiments, Used Tools and Learning Outcomes

Used tools	Learning outcomes (LOs)	
	Cognitive LOs	Psychomotor LOs
Mobile application: "WiFi Scanner","WiFi Analyzer" and "Speed test"	Understanding mechanisms of radio wave propagation	Using mobile devices as a measurement tool for WiFi networks
	Knowing the differences between indoor/outdoor radio wave propagation	
	Understanding interference problems	
Mobile application: "OpenSignal Maps", "GSM signal monitor", "G-Net Track" and "Speed test"	Understanding concepts related to cellular telephony, such as roaming and handover	Using mobile devices as measurement tools for Mobile networks
	Understanding mechanisms for propagation of radio waves	
	Knowing the differences between indoor/outdoor radio wave propagation	
	Understanding Interference problems	

3.2.A. Measurement of a WiFi Network

The objective of this experiment is to deeply analyze a WiFi Network using the student's own mobile terminal or tablet device. The students are asked to prepare a full study of the WiFi networks available in the University Campus (one or two buildings), or any other location of their choice. This study consists of complete coverage maps containing signal strength data and data rate values, along with a detailed explanation of the data acquired. They must use free mobile applications, such as "WiFi Scanner & Net Discovery" [16] or "WiFi Analyzer" [17] for extracting information, or "Speed Test" [18] to measure data rates, using their devices as the measuring equipment in both cases. The students achieve the psychomotor LOs "Use with mobile devices as measurement tools for WiFi Networks" when providing the coverage maps of the WiFi Network with correct signal strength values in the report. Some screenshots of these applications are shown in Figure 4.

3.2.B. Measurement of Mobile Networks

Simirlaly as in the case of WiFi networks, the students are required to study and analyze different mobile networks. The students must search for applications that allow their mobile devices to provide them information regarding the mobile network. The applications most frequently used are "Open Signal Maps" [19], "GSM signal monitor" [20] and "G-Net Track" [21], which are available without costs in Google play store. These applications provide information about signal strength, direction of arrival and RSSI, among other mobile network parameters. Moreover, some of these applications are also capable of providing the data-rate of the networks and locating the different mobile stations on a map. Figures 5 to 8 show some screenshots of different mobile applications. With the "GSM Signal Monitoring" application, we can see the signal level that we are receiving and, as it is seen in Figure 5, when the handover occurs, produced by the movement of the user, that is, as the mobile passes, it jumps from one station to the other. In Figure 6, the application "Open Signal" places us the different base stations on a map and it performs measurements of velocity. The "G-Net Track" application indicates in different colors the signal level or station to which the mobile is connected, this application is very appropriate to try to make coverage maps, the screenshots are shown in Figure 7. Finally the "Sensorly" application has been used to perform speed tests, connecting to different base stations with different standards of wireless networks. In Figure 8 we can see the difference in speed in data transmission, when it is connected to a WiFi network, a HSDPA network or an EDGE network. Once the students are familiar with those applications, they should follow a route through the city identifying the different mobile phone networks available on their route, and different network parameters. They must also think about the various phenomena associated with signal propagation: they must find the approximate point where the handover occurs, for instance. They are also asked to make a small sketch including the mobile network cells and the different mobile systems available along the route. With this activity, the cognitive LOs "Understanding concepts related to cellular telephony, such as roaming and handover" is achieved, providing a report which contains examples of handovers produced on the student's device.

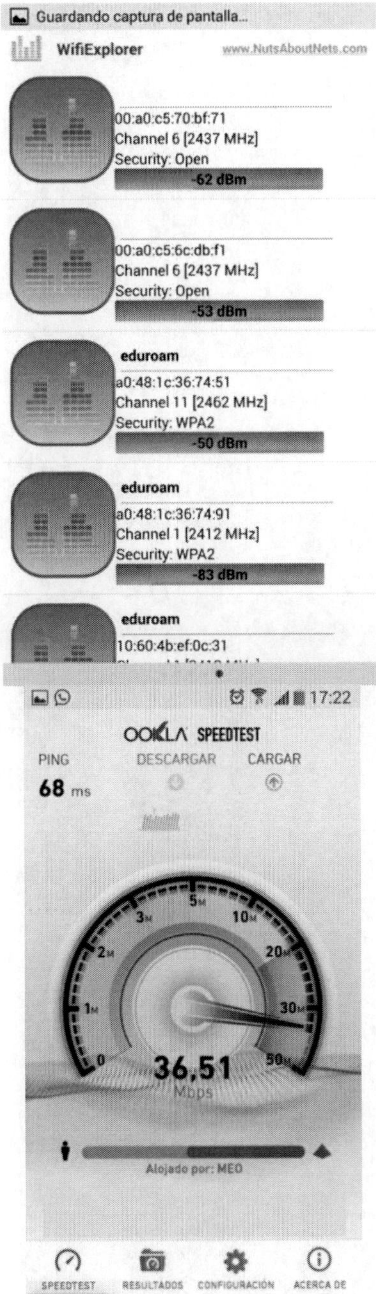

Figure 4. Screenshots of Wi-Fi Explorer Scanner & Net Discovery and WiFi Analyzer applications.

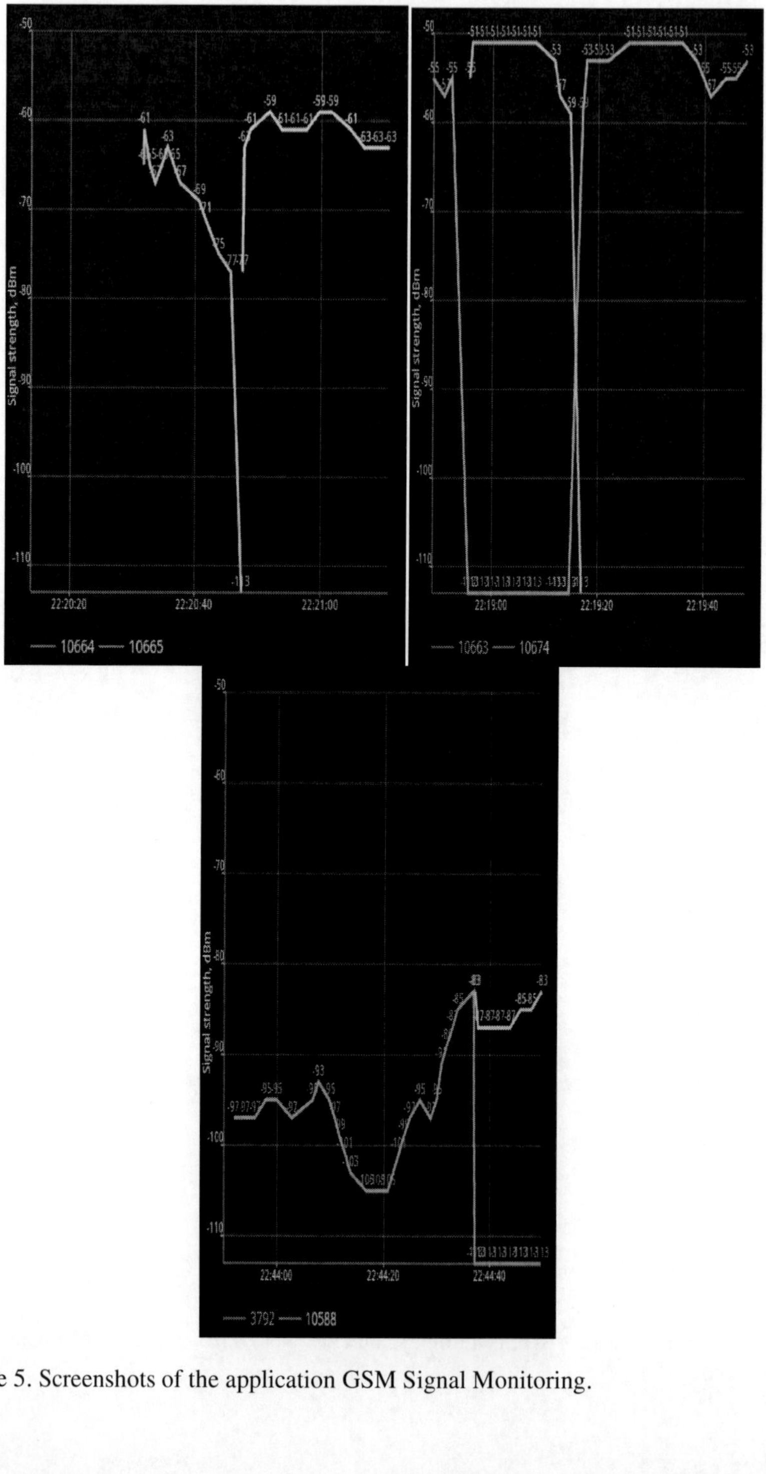

Figure 5. Screenshots of the application GSM Signal Monitoring.

Figure 6. Screenshots of the application Open Signal Maps.

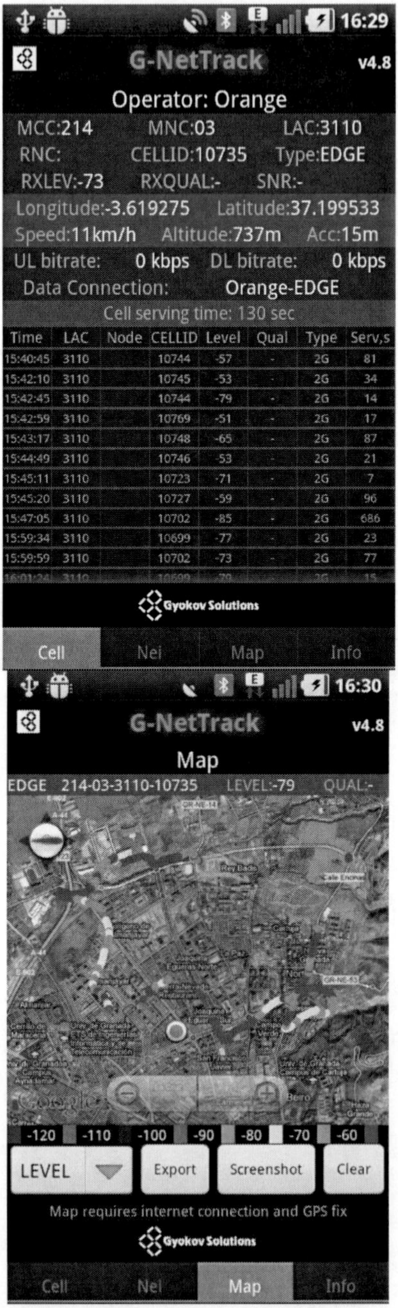

Figure 7. Screenshot of the application G-Net Track.

Figure 8. Screenshots of the application Sensorly.

Table III. Skills to Be Acquired

Skill to be acquired	Level of skills development (1 minimum - 5 maximum)
Self learning	4
Creative Thinking	3
Taking initiative	5
Development of critical thinking	4
Ability to work in group	4
Communication Skills	3
Troubleshooting	3
Data Analysis	4
Oral Presentations	4
Reporting	5

3.3. Results

3.3.A. Learning Outcomes Achieved

The achievement of the learning objectives is determined in two ways: first, with the evaluations of the submitted reports, then, with the defense and personal knowledge evaluation. The results were very satisfactory since over 85% of students demonstrated to have acquired more than 50% of the learning objectives. The results of the evaluation of the post-experiment reports and the final exam are shown in Figures 9 and 10. The final exam is divided into 2 different parts that correspond to the 2 different laboratory experiments. The report scores are higher than those achieved in the exam. In Figure 9, it is shown the mean and standard deviation scores in the final exam and post-experiments reports from different academic years (Response 0 to 10). Figure 10 shows the mean and standard deviation scores in the final exam and post-experiments reports from different experiments (Response 0 to 10).

The results of the final exam lead to the following conclusions:

- The results in both years are reasonably good.
- Post-Report 1shows the maximum scores.
- Test exam for Experiment 2 are slightly better.

Table IV. Questions

N.	Question
1	The teacher promotes the student motivation with the contents and practical activities
2	The teacher provides guidance for student self-learning
3	The assistance received from the teacher is effective for learning
4	The assistance from the teacher is effective for learning
5	In general, the teacher has successfully completed his teaching

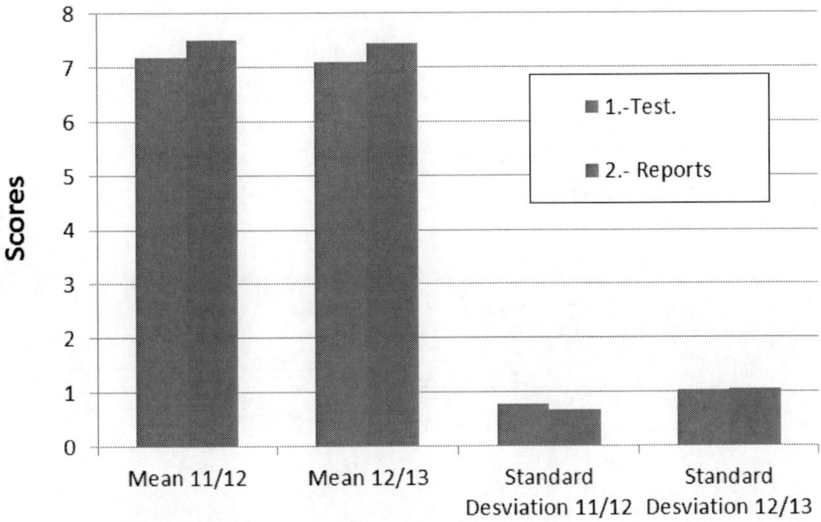

Figure 9. Mean and standard deviation scores in the final exam and post-experiments reports from different academic years (Response 0 to 10).

4. Student's Opinion on Both Projects

4.1. Students' Opinion of the Project for Project I

This section shows the opinion of the students about the laboratory course and has been obtained through anonymous surveys. The survey was given to the students before the final exam. It was composed of 5 different questions shown in Table IV. Figure 11 shows the mean values for the Radio navigation course and the results are compared with the results of the mean of all the subjects in the last year. The responses are ranged from 0 (Very poor) to 5 (excellent).

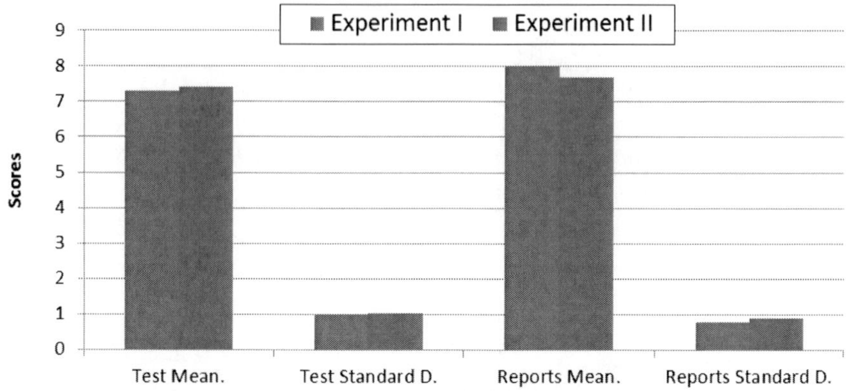

Figure 10. Mean and standard deviation scores in the final exam and post-experiments reports from different experiments (Response 0 to 10).

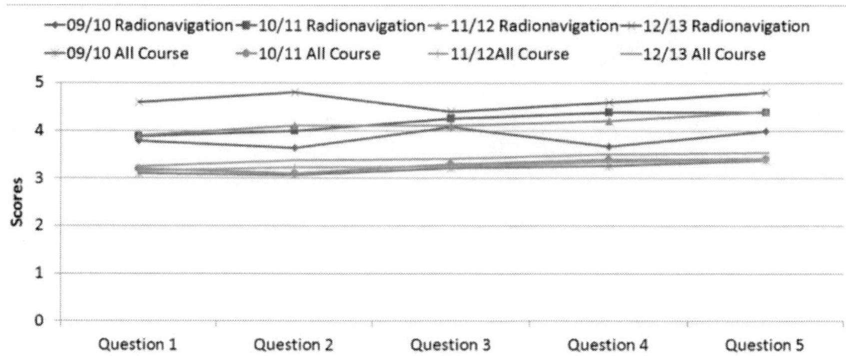

Figure 11. Laboratory scores.

From the results of the survey, the opinion on the employed methodology is very good, almost every item scores over 3.5 on 5 and the support of the teachers during the project has been considered very positively.

4.2. Students' Opinion of the Project for Project II

This section provides the opinion of the students regarding the project II, again, through anonymous surveys. The survey shows the degree of satisfaction with the projects, in each item there are several questions and Figure 12 shows the mean values for all students and questions by item.

From the results of the survey, the following can be concluded: first, a good score in all items was achieved and second, the students valued the experience and the methodology used more than the skills learned. Finally, the support of teachers during the project was evaluated very positively by the students in spite of the reduced number of face to face lessons.

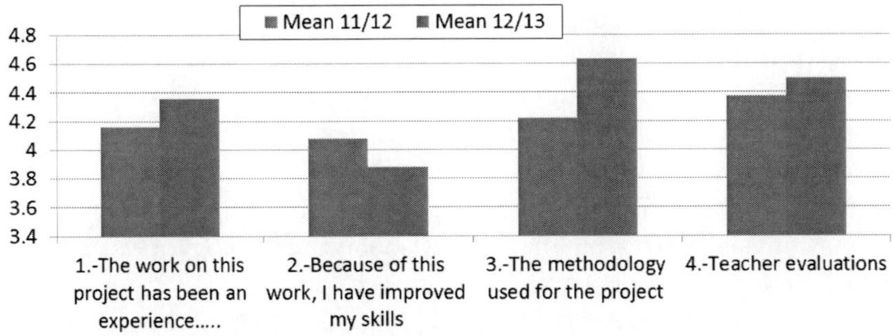

Figure 12. Students' opinion.

CONCLUSION

First, we have presented a literature review of the main works of PBL in issues related to m-learning and u-learning. Next, the authors present two implementation experiences of two real cases m-learning and u-learning. In the first experience, it is presented a small experience with a radio navigation application for smart phone similar to some software developed for computers. However in the second experience the mobiles are used as measurement tools. The laboratory course described offers three main advantages in relation to the laboratories used by other universities: first, it provides ubiquitous learning, including measurements which are not typically considered in this kind of ubiquitous learning; second, the material needed to implement the laboratory work has no cost except for the student terminal; third, the students learn how to exploit the capabilities of their mobile devices. Results show that most of the objectives have been achieved. These educational experiences have been very positive for the level of satisfaction and involvement of all parties. Other universities may establish similar experiences, taking advance of the projects outlined in this work.

REFERENCES

[1] Liaw S. S., Hatala M. and Huang H. M., Investigating acceptance toward mobile learning to assist individual knowledge management: Based on activity theory approach, *Computers & Education*, vol. 54. No. 2, pp. 446-454, 2010.

[2] Evans C., The effectiveness of m-learning in the form of podcast revision lectures in higher education, *Computers & education*, vol. 50, no. 2, pp. 491-498, 2008.

[3] Marwan M. E., Madar A. R. and Fuad N. An overview of mobile application in learning for student of Kolej Poly-Tech Mara (KPTM) by using mobile phone, *Journal of Asian Scientific Research,* vol. 3 no. 6, pp. 527-537, 2013.

[4] Martin F. and Ertzberger J., Here and now mobile learning: An experimental study on the use of mobile technology *Computers & Education* vol. 68 pp. 76-85 2013.

[5] Jones A. C., Scanlon E., Clough G., Mobile learning: Two case studies of supporting inquiry learning in informal and semiformal settings, *Computers & Education* Vol. 61, pp 21-32, 2013.

[6] Abachi Hamid R., Muhammad Ghulam, The impact of m-learning technology on students and educators, *Computers in Human Behavior*, Vol. 30, pp. 491-496 January 2014.

[7] Furió D., González-Gancedo S., Juan M.-C., Seguí I. and Costa M., The effects of the size and weight of a mobile device on an educational game, *Computers & Education*, Vol. 64, pp 24-41, May 2013.

[8] Molina A. I., Redondo M. A., Lacave C., Ortega M., Assessing the effectiveness of new devices for accessing learning materials: An empirical analysis based on eye tracking and learner subjective perception, *Computers in Human Behavior*, Vol. 31, pp. 475-490, February 2014.

[9] Yang G., Chen N.-S., Kinshuk, Sutinen E., Anderson T. and Wen D., The effectiveness of automatic text summarization in mobile learning contexts, *Computers & Education*, Vol. 68, pp 233-243 October 2013.

[10] Wu W.-H., Jim Wu Y.-C., Chen C.-Y., Kao H.-Y., Lin C.-H. and Huang S.-H., Review of trends from mobile learning studies: A meta-analysis, *Computers & Education*, Vol. 59, pp 817-827, September 2012.

[11] Chen C.-C., Huang T.-C., Learning in a u-Museum: Developing a context-aware ubiquitous learning environment, *Computers & Education*, Vol. 59, pp. 873-883. November 2012.

[12] Hwang, G. J., Yang, T. C., Tsai, C. C., & Yang, S. J. H. (2009). A context-aware ubiquitous learning environment for conducting complex experimental procedures. *Computers & Education*, 53(2), 402-413.

[13] Chiou, C. K., Tseng, J. C. R., Hwang, G. J., & Heller, S. (2010). An adaptive navigation support system for conducting context-aware ubiquitous learning in museums. *Computers & Education,* 55, 834-845.

[14] *Nav Trainer Basic* https://play.google.com/store/apps/details?id=com.fleminger.navtrainerBasic&hl=es (Accessed: 10 Feb., 2014).

[15] *Nav Trainer Pro for Pilots* https://play.google.com/store/ apps/details?id=com.pilotscafe.apps.navtrainer&hl=es (Accessed: 10 Feb., 2014).

[16] *Wifi Scanner & Net Discovery,* (Online). Available: https://play.google.com/store/apps/details?id=com.nutsaboutnets.wifiexplorer (Accessed: 5 Sep., 2014).

[17] *WiFi Analyzer,* (Online). Available: https://play.google. com/store/apps/details?id=com.farproc.wifi.analyzer (Accessed: 5 Sep., 2014).

[18] *Speed test,* (Online). Available: https://play. google.com/store/apps/details?id=org.zwanoo.android.speedtest l (Accessed: 5 Sep., 2014).

[19] *Open Signal-mapas 3G/4G/WiFi* (Online). Available: https://play.google.com/store/apps/details?id=com.staircase3.opensignal (Accessed: 5 Sep., 2014).

[20] *GSM Signal Monitoring,* (Online). Available: https://play.google.com/store/apps/details?id=com.signalmonitoring.gsmsignalmonitoring 1 (Accessed: 5 Sep., 2014).

[21] *G-Net Track,* (Online). Available: https://play. google.com/store/apps/details?id=com.gyokovsolutions. gnettrack&hl= es (Accessed: 5 Sep., 2014).

In: Project Based Learning on Engineering
Editors: J. Valenzuela-Valdés et al.
ISBN: 978-1-63482-217-6

Chapter 9

CONCLUSION, IMPLICATIONS AND FUTURE

Pedro J. Pardo*[*] *and Juan F. Valenzuela-Valdés
Centro Universitario de Mérida, Universidad de Extremadura,
Mérida, Spain

ABSTRACT

This chapter sets out to establish, as conclusions, the major contributions made in this joint work dedicated to projects based learning applied to engineering. It also aims to provide a critical overview of the current system of university education and those that are, in our view, future works on this subject line.

CONCLUSION

After reading this book, we believe it is clear that the use of methodologies of problem-based learning in general, and project-based learning applied to engineering particularly, is worthwhile for several reasons. On the one hand it provides greater achievement of academic goals set by the professor as compared to those obtained with classical teaching. Especially meaningful is the topic regarding certain skills such as teamwork, leadership and oral communication. These results are not new as there are clear examples

[*] Emails: pjpardo@unex.es; juanvalenzuela@unex.es

where the use of innovative teaching methods involves better student outcomes [1], but the results are new in terms of the application to engineering. This branch of knowledge traditionally suffers from innovative methodologies that facilitate the acquisition of new competencies that the European framework for higher education is calling on all bachelor degrees [2], including engineering [3].

From an educational point of view, this work has shown a review of the state-of-the-art of problem-based learning as well as a study of development of scientific publications on this topic in recent years. The fundamental aspects of PBL were covered and an implementation methodology was proposed. The evaluation topic was introduced by rubrics. All of these make this book a compendium of PBL applied to engineering.

Moreover, surveys of students who have participated in these innovative experiences indicate their satisfaction with the new teaching methods employed. They especially appreciate the extra motivation generated for them in participating in this type of collaborative experience, groupware and outside the traditional classroom.

Another point to consider as a remarkable conclusion of this work is the demonstration that these types of collaborative methodologies are not linked to a unique learning environment. There have been examples of successful application of the method of project-based learning (PBL) using asynchronous online learning platforms, synchronous virtual classroom, and examples of mobile-learning and ubiquitous-learning. All of these platforms and learning environments are valid not only for the case of PBL. They can be used to solve the problem of group or individualized tutoring or mentoring. In any case, the degree of student satisfaction has been very high.

IMPLICATIONS AND FUTURE

The results of the experiences based on PBL presented throughout this collective work support the use of this methodology as a complement to traditional teaching in the classroom. However, the learning environments and tools described can be used in semi-face or virtual education. Until now, this type of virtual education was reserved for minor cases of post-graduate studies or training courses. Only very few universities were specialized in distance learning, and virtual education was a typical teaching method. These universities, in most cases, have a physical network of mentors to whom the pupils can contact in case of doubt.

However, our society is changing: a high degree of competitiveness and excellence are looked for by companies and employers. In addition, the technological development we are experiencing as a fact of our real lives makes traditional classroom education lose weight off the virtual teaching. Until a few years ago, the vast majority of students who enrolled in any bachelor degree corresponded to young people between 18 and 21 years old without dependents. However the student profile in recent years has changed. The economic crisis has made many people who have lost their jobs or have never been working, decide to recycle or complete their education [4]. In many cases, the social and familiar situation of these students is very different from the previous students. These students are looking for the best way to reconcile their life with studies. All this means that, even in the model of presencial education, the use of these technologies facilitate studies for this type of part-time-student or many other situations (company, family, etc.). Traditionally, it is on courses related with technical skills, very common in engineering, where you have the greatest impediment to apply this type of virtual or semi-face teaching. However, the results shown in this paper indicate that it is possible, adapting in each case of needs, to use the PBL methodology to teach technical skills in this type of student. We consider the study, exploration and adequacy of all learning environments used in this work to solve these problems related to the development of techniques in teaching competences, which opens virtual engineering a workline for the immediate future. Logically, the concept of engineering is very broad and diverse, but previous results encourage starting this way.

It remains, however, the question of how the academic authorities will consider these types of teaching methods and how they will compute their use in the workload of teachers. In an educational system like the European university system, and especially the Spanish, where tradition sometimes gets confused with immobility, the efforts of teachers applying these methodologies should be considered.

REFERENCES

[1] Martínez, G., Pérez A. L., Suero M. I., Pardo P. J. (2012) "The Effectiveness of Concept Maps in Teaching Physics Concepts Applied to Engineering Education: Experimental Comparison of the Amount of Learning Achieved With and Without Concept Maps," *J. Sci. Educ. and Technol.*, 5, 1-11.

[2] The Bologna Declaration on the European Space for Higher Education: An explanation, 1999, http://www.nvao.net/page/downloads/Bologna_Declaration.pdf, accessed 19 November 2014

[3] The Global Engineer, Incorporating global skills within UK higher education of engineers, 2208, http://eprints.ioe.ac.uk/839/1/Bourn 2008Engineers.pdf, accessed 14 December 2014

[4] The Social Crisis Behind the Economic Crisis – the Millions of Young People Unemployed, ITUC REPORT May 2012 • 2, ITUC, Belgium.

INDEX

A

B

C

F

G

H

I

J

L